Philosophy, technology, and
the environment /

DATE DUE

			PRINTED IN U.S.A.

Philosophy, Technology, and the Environment

edited by David M. Kaplan

The MIT Press
Cambridge, Massachusetts
London, England

Set in Stone Sans and Stone Serif by Toppan Best-set Premedia Limited. Printed on recycled paper and bound in the United States of America.

Library of Congress Cataloging-in-Publication Data

Names: Kaplan, David M., editor.
Title: Philosophy, technology, and the environment / edited by David M. Kaplan.
Description: Cambridge, MA : The MIT Press, 2017. | Includes bibliographical references and index.
Identifiers: LCCN 2016026442 | ISBN 9780262035668 (hardcover : alk. paper)—9780262533164 (paperback)
Subjects: LCSH: Technology--Philosophy. | Technology--Environmental aspects. | Sustainable engineering. | Environmental ethics.
Classification: LCC T14 .P5424 2017 | DDC 601--dc23 LC record available at https://lccn.loc.gov/2016026442

10 9 8 7 6 5 4 3 2 1

My father-in-law, Robert Schreiber, always read whatever I sent him.
I think he would have really liked this book.

Contents

Acknowledgment

I owe a special debt of gratitude to Philip Brey for organizing and supporting the workshops that led to the publication of this edited volume. This project never would have happened without his work and assistance.

Introduction

David M. Kaplan

Environmental philosophy and philosophy of technology have a lot in common. Both fields explore the positive and negative aspects of human modifications of the world. Both question the limits of technology in relation to natural environments, animals, plants, and food. Both examine if human making and doing is compatible with nature or wholly different from it. And both examine the difference between what is considered to be natural and artificial. Technology and the environment further intersect in a number of issues, such as climate change, sustainability, geo-engineering, and agriculture. The reason for the overlap is fundamental: Environmental issues inevitably involve technology, and technologies inevitably have environmental impacts. Technology and the environment are like two sides of the same coin: Each is fully understood only in relation to the other. Yet, despite the ample overlap of questions concerning technology and the environment, the two philosophical fields have developed in relative isolation from each other. Even when philosophers in each field address themselves to similar concerns, the research tends to be parallel rather than intersecting, and the literatures remain foreign to one another. These divergent paths are unfortunate. Philosophers from each field have a lot to contribute to the other.

The aim of this book is to show how technology is central to environmental philosophy and the environment is central to philosophy of technology. It aims to broaden and deepen philosophical discourse on technology and the environment in order to show how intimately related they are. By combining the philosophic expertise of two perspectives, this volume will attempt to "green" philosophy of technology and to "technologize" environmental philosophy. The potentially new hybrid discourse might help to start new dialogues not only between these two areas of philosophy but also among other disciplines and broader, non-academic publics.

This is not to suggest that philosophers of technology have heretofore never addressed the environment, or that environmental philosophers have never addressed technology. They have—just not as much as you might expect. A survey of the two leading journals in philosophy of technology and environmental philosophy is telling. *Research in Philosophy and Technology* (founded in 1978 and renamed *Techné* in 2000) is the oldest and historically most important journal in its field. *Environmental Ethics* (founded in 1979) is the oldest and most historically most important journal in its field. Over the last 35 years *RPT/Techné* published 56 articles that examine some aspect of nature or the environment. *Environmental Ethics* published 29 articles that examine some aspect of technology over the same stretch.[1]

In 1992 and 1999 *RPT/Techné* devoted two entire volumes to the relationship of technology to the environment to attempt a rapprochement between the two fields. The primary aim of the second volume, according to guest editor Marina Banchetti, is "the possibility, perhaps even the desirability, of merging environmental philosophy with philosophy of technology" (Banchetti 1999, 4). Though the two specializations are not ignorant of the other, on her reckoning "participants in neither subdiscipline are anywhere near being sufficiently prepared in the discussion and insights of the other." Environmental ethics overemphasizes wilderness and views human technological activity negatively. Philosophy of technology displays a "naïve anthropocentrism" by focusing the role of devices and machines on social, political, and economic affairs to the exclusion of ecological concerns. The volume did not, of course, achieve Banchetti's hope of forming "one common focus and scope" between the two specializations (ibid., 5).

The articles in *RPT/Techné* on the environment since 1978 fall into five thematic constellations:

- the historical development of the concept of nature in antiquity, modernity, and non-Western philosophical traditions
- the relationship of technology and nature, mostly from the two special editions in 1992 and 1999 devoted to the environment
- the relationship of science and nature, which typically denounce scientific reductivism and either call for a broader conception of scientific practice or affirm non-reductivist theoretical frameworks
- two articles on religion, technology, and nature that trace humanity's place in nature to Biblical origins

• several articles on nature and technology in politics and public policy that range from questions of political rights and public representation in technical affairs, to the relationship among technologies, regulations, and ecological sustainability (Miller 2015, 34)

The articles in *Environmental Ethics* on technology fall into four thematic areas:

• the historical development of the concept of technology and technological rationality in antiquity, Modernity, and non-Western philosophical traditions
• human making (taken broadly) and the environment, including the restoration ecology debates in the mid 1990s about whether we can intervene at all in the natural world without also turning it into an artifact
• the relationship of science, technology and nature, many of which refer to the works of Martin Heidegger and Albert Borgmann
• religion, technology, and nature and the connection between Lynn White Jr.'s interpretation of Medieval Christianity and environmental degradation (Miller 2015, 76)

Figure I.1 is a chart of cross-specialization articles in *RPT/Techné* and *Environmental Ethics* since 1979 (Miller 2015). We see an increased concern for the environment in *RPT/Techné* in the 1990s, mainly due to the two special editions and the restoration ecology debates. We also see a relative lack of interest in the environment in every decade except the 1990s. The spike in interest given to nature in *RPT/Techné* does not have a counterpart in articles on technology in *Environmental Ethics*. Instead we see, on average, one article per year for the last the 35 years.

This volume hopes to return to the spirit of the 1990s by bringing together philosophers to address the intersection of technology and environment. Some of the contributors belong squarely to one specialization, some work in both, and some belong to neither. But they all share the intuition that technology and the environment intersect in ways that are philosophically relevant and practically important. The contributors share a sophisticated understanding of man-made things as more than value-free technical devices but rather as complexes of materials, skills, purposes, and relations. Man-made things are best understood in context, relative to particulars uses and users, in relation to varied social and environmental settings. Each chapter addresses how the technology-environment relationship has been conceived and how we might understand it better. The authors wrestle with the question about the limits of legitimate technological interventions in the natural world, what it means to say that a concern for the

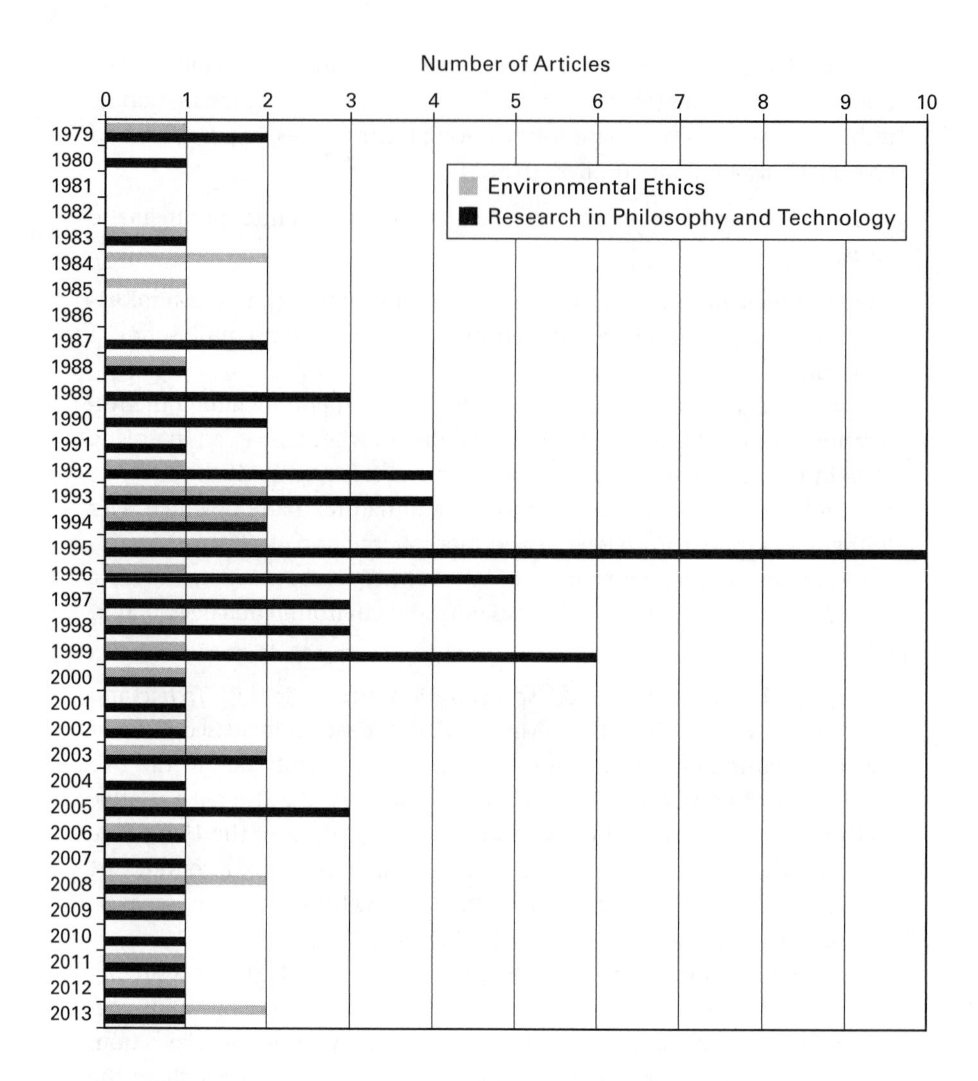

Figure I.1
Cross-specialization articles.

environment can be designed into things, how consumerism relates us to artifacts and environments, and how food and animal agriculture raise questions concerning both culture and nature. Although this volume might not develop truly hybrid discourse, the hope is that it will join and even advance the philosophical conversation about how things can and should relate to the natural world.

In the first chapter, J. Baird Callicott, one of the founders of environmental philosophy, questions the basic premise of Lynn White Jr.'s seminal essay "The Historical Roots of our Ecological Crisis." White famously attributes the environmental harms in industrialized nations to a few lines in Genesis in which God creates man in his image, gives man dominion over the rest of creation, and commands him to subdue the Earth. Callicott's chapter questions not so much the narrow reading of the Bible (or Torah) but rather White's very epistemic assumption: that what we do depends on what we think. The fundamental premise of 1980s environmental philosophy and the goal of every politically minded environmental philosopher was to uncover underlying metaphysical and epistemology presuppositions of thought and action in order to promote preferable alternatives (either non-Western or overlooked Western traditions). On this reckoning, we need to rethink the nature of nature, human nature, and the relationship between humans and nature as the first and most important step toward saving the world from ecological disaster.

But Callicott reminds us that the Lynn White Jr. of *Medieval Technology and Social Change* (1962) also proposes a theory of technological determinism to explain the fate of the West that flies in the face of the intellectual determinism of White's essay "Historical Roots." Ordinary objects, like White's stirrup example, bring about massive social upheavals. So which is it? Is the mechanistic worldview of Descartes and Newton the product of Christian theology or mechanical technologies? Is what we do dependent on what we think, or is it the other way around? Perhaps the nature of nature, human nature, and the relationship between humans and nature are more affected by things than by ideas. If so, environmental philosophers have to give up the pretense that they alone can save the world from environmental destruction because they alone are expert at uncovering underlying conceptual presuppositions. Revolutionary developments in real material things are just as important as revolutionary ideas.

The next chapter is by Don Ihde, one of the founders of philosophy of technology. Ihde addresses the "congenital dystopianism" shared by environmentalists, environmental philosophers, and philosophers of technology. Each group employs a "rhetoric of alarm" that connects the use of technologies with environmental degradations. The godfathers of philosophy of technology, Karl Marx, Jacques Ellul, and Martin Heidegger warn that a technology-driven society will produce alienation, conformity, and dehumanization. Similarly, the literature in environmental philosophy is filled with gloomy forecasts of overpopulation, food shortages, and irreversible climate change. Ihde wishes neither to deny the reality of twenty-first-century ecological challenges nor to advocate for an unrealistic technological utopia. Instead he calls attention to how "excessive rhetorical strategies" have locked us into a false dichotomy: either technological-environmental utopianism or dystopianism.

Ihde also warns of "microsolutions to macroproblems"—thinking globally but acting locally. Although there is nothing wrong with small-scale responses to big issues (recycling, banning toxins), they often flirt with "nostalgic romanticism." The problem is that we have not yet figured out what our technologies can or should do, or what the environmental crises actually are. So long as we continue to accept either utopian or dystopian forecasts, we are unlikely to bring either technologies or ecosystems into appropriate focus. Techno-environmental problems are complex, ambiguous, and interwoven; they rarely lend themselves either to an easy technofix or a simple solution. The hardest problem of all is how to turn major actors in the economy (large development projects and multinational corporations) "green." The challenge for a proactive philosopher, says Ihde, is to get in on the ground floor of technological research and development in order to help figure out how to green the economy itself.

Kyle Powys Whyte, Ryan Gunderson, and Brett Clark examine the notion of insidiousness in the philosophy of technology and environmental philosophy—the idea that the adoption of a new technology erodes intimate relationships with the environment. The alleged erosion undermines not only ecological relationships but also the social and cultural relationships that sustained ways of life that were less dependent on complex technologies. Whyte, Gunderson, and Clark begin their chapter by rejecting the idea that technologies are neutral or disinterested but instead embody values, preferences, and lifestyles. They then examine several versions of technological non-neutrality found in environmental philosophy and sociology, in traditional ecological knowledge, and among political economists. These schools of thought call attention to the social, economic, and political contexts that influence technological innovation and administration and call into question the technological optimism found in the environmental mainstream. Whyte et al. argue that technologies can indeed sometimes transform our social and environmental relationships in insidious ways. They conclude with five questions philosophers and social scientists should consider when examining different technology-society-environment interfaces:

- What social and social-environmental relations give rise a technology, and how does it reproduce or transform these relations?
- Who is in control of the design and application of the technology?
- How is the technology governed? What social groups and environments benefit from it?
- What social groups and environments are harmed by the technology?
- What values, interests, and politics are reflected in the design and application of the given technology?

The answers to these questions may not determine that technology is insidious, but they are likely to counter the view that it is neutral and disinterested.

In the next chapter, Paul B. Thompson takes a more optimistic view of the ways that technologies interact with their environments. He analyzes recent opposition to novel technologies and finds several patterns. Whereas earlier generations of critics interpreted the impact of new technologies in largely social terms, present-day critics focus on risks to health, to biodiversity, and to the integrity of habitats and ecosystems. Yet defenders and critics share the same fundamental assumption that technological innovation is the main source of greater efficiency in production processes. Although they question the ways that social institutions incentivize innovation and distribute benefits, innovation as such is always seen as a good thing (think renewable energy or sustainable development)—except when it comes to certain emerging technologies: agricultural biotechnologies, synthetic biology, and nanotechnology. Then public perception is skeptical, negative, even outraged. The reason is not entirely clear. We give a pass to information and medical technologies. Why are some emerging technologies seen as insidious and others not?

Thompson turns to risk assessment to figure out what makes some technologies more disturbing than others, specifically why some risks are considered to acceptable and others are rejected. He examines the "social amplification of risk," the cognitive and social phenomena that distort perception and cause people sometimes to see a situation as more risky than it is and sometimes to see it as less risky. Thompson identifies two different approaches to risk amplification: purification and hybridization. The former excludes irrational social fears, outrage, and distrust from a risk assessment; the latter takes these motivating influences seriously and incorporates them into a risk assessment. Thompson warns that purification can engender the suspicion that powerful actors are indifferent to social perceptions, and suggests that hybridization can be an effective response to the perception of environmental harms. He attempts to steer a path between the two approaches in order to balance scientific risk management with effective responses to social injustices.

Each of the first four chapters questions the underlying pessimism in early attempts to relate technology to the environment. Callicott and Ihde find the pessimism unwarranted; Whyte, Gunderson, and Clark find it sometimes and conditionally warranted; Thompson explains why it is sometimes warranted. The remaining chapters focus less on theoretical issues about the technology-environment relationship itself and more on how in fact they do (and ought to) relate. Although no less philosophical than the first four chapters, these chapters examine actual intersections of technology and environment. They fall under four themes: environmental remediation, sustainable design, consumption, and agriculture. Each issue is a conundrum for our philosophical understanding of both technology and the environment. The authors, however, eschew either overly gloomy or dopily optimistic assessments of the environment. Instead they find a complex nexus of mutual implication between humans, artifacts, and nature.

Benjamin Hale lays out the arguments for morally permissible environmental remediation. He first distinguishes among several kinds of remediation (carbon sequestration, ocean fertilization, pollution cleanup, and atmospheric scrubbing) in order to distinguish the question of *whether* remediation is permissible from the question of *which* remediations are permissible. Hale's argument proceeds in four stages. First he constructs an elaborate scenario involving a town with three companies and their emissions in order to challenge the intuition about what the wrong of pollution consists in. Second, he argues that any proposed remediation should be evaluated not only in terms of its consequences but also by the motives and antecedent conditions that make a remediation an option in the first place. Third, he constructs scenarios to argue that we are obligated to act only over areas for which we are reasonably responsible. Finally, he argues that a permissible remediation may use only technologies that return the world to how it was before and not technologies that result in a new state of affairs. In other words, we can legitimately undo harms but we can't legitimately patch over them. The primary moral consideration of a remediation is whether it respects the experiences, the interests, and the ends of others. Hale argues that we should remove pollutants that we have directly contributed to but that we should not remediate pollutants with technologies that add something new to the environment—unless all affected parties could or would assent to the technological remediation.

Clare Heyward, Steve Rayner, and Julian Savulescu focus more narrowly on geo-engineering remediations, such as carbon-dioxide removal and solar radiation management designed to reduce human-caused climate change. In their chapter, they examine the legitimacy and social control over the research, development, and any eventual deployment of geo-engineering. They argue that geo-engineering is permissible in principle but that all geo-engineering R&D should be subject to some sort of governance given its potential to affect everyone in the world. They defend the Oxford Principles of ethical-political decision-making: that geo-engineering is in the public interest and should be regulated as a public good, that there should be public participation in geo-engineering decision making, that geo-engineering research should be transparent and available to the public, that risk assessments should be conducted by independent bodies and be directed toward both the environmental and socioeconomic impacts of research and deployment, and that the legal, social, and ethical implications of geo-engineering should be addressed *before* a project is undertaken or technology deployed. Heyward et al. then compare the Oxford Principles favorably with the three main alternative models that guide geoengineering development, arguing that it has a wider scope of application than the alternatives and that they lend themselves to action-guiding recommendations and regulations appropriate to different technologies while preserving long-standing environmental and political values.

The next three chapters are on design and sustainability. The common thread among them is the insight that sustainability is achieved only if designed into a technological product or process. In other words, it has to be built into the life cycle of something man-made. Although this insight might seem trivial, it speaks to the intimate connection between technology and the environment. Concern for the welfare of the environment is built into things; in turn, sustainability is made possible thanks to man-made things. If this insight is correct, it can help move us beyond the ideas that technology and the environment are opposed to one another—one social, one natural. Instead, the notion of sustainable design builds a theoretical and practical bridge between the things people make and the environments that are independent of us. We may legitimately manage nature thanks to sustainable design.

In his chapter, Ibo van de Poel claims that there is substantial disagreement about the exact conception of sustainability even if there is general agreement on its desirability. This disagreement will not be resolved easily, because it represents fundamentally conflicting normative views on society and nature. Van de Poel argues that we should see sustainability as a "compounded" value (not a mere technical issue) that consists of intergenerational justice, intragenerational justice, and care for nature. These values often come into conflict. Furthermore, the meaning of "sustainability" varies according to different design projects. The three most common ways of dealing with value conflicts are *life-cycle analysis* (which compares multiple environmental impacts and aggregates them into one measure), *respecification* (which identifies higher-order and less controversial values to be specified into a design), and *innovation* (which develops new options that meet all design requirements). Value conflicts usually can be handled by specifications of shared values or by new innovations. But if none of those options works, van de Poel recommends a "values hierarchy" that organizes conflicting interpretations of a specific project in order to determine the available design options. He proposes "value dams" and "value flows" to manage conflicts among stakeholders who may have different conceptions of sustainability. A value dam would prevent adoption of design features that are strongly opposed by at least one or more stakeholders, and a value flow would promote design requirements that fit a number of different conceptions of sustainability. Just because sustainability is a contested concept and stakeholders have conflicting values does not mean we cannot attempt to find specific design requirements for sustainability that can guide the development of new, greener technologies.

Next, Braden Allenby examines industrial ecology, a field of study devoted to the relationships among industrial, economic, and natural systems. It analyzes environmental impacts of products and processes associated with industrial systems in order to identify sustainable development strategies. At the heart of industrial ecology is a biological analogy—specifically the idea that ecosystems tend to use materials and energy efficiently because they evolved to do so and that, therefore, industrial systems should attempt to emulate them. Allenby explains the origins of environmental design and how it can be incorporated into all aspects of engineering design for sustainability. Industrial ecology is of interest not just because of the suite of methods and tools that it encompasses but also because it defends a sustainable vision of how technological activity relates to the larger ecosystems.

Yet Allenby also identifies several weaknesses of environmental design. To begin with, the development and implementation of industrial ecology has been too fixed and unchanging; it has not reflected evolving conceptions of either environmentalism or the globalization of the economy. Also, its focus has been too limited, paying too much attention to manufacturing and materials and not enough to services or to information and communication technologies. Worse, the focus on high tech and developed economies has prevented industrial ecology from having much of an effect in developing nations. But even more pressing is that industrial ecology has not modified its conception of the environment in light of increasing human transformations on the natural world. If we are living in the anthropocene, the biological model of the environment is not only dated but woefully inadequate to account for wide range of human values that should be incorporated into design. Human systems are far more complex than biological systems. We should be careful about relying too much on limited biological and ecological metaphors. Allenby, however, remains optimistic that industrial design can overcome its limitations, evolve alongside the environmental movement, and offer viable alternatives to theorists and practitioners concerned with sustainability.

Zhang Wei examines the role of design in environmentally sustainable products. He views the environmental crisis as related to unsustainable product design. Most models of sustainable design focus too much on the environmental impacts of products and too little on their interactions with users. The result can be paradoxical consequences in which undesired effects can offset any environmental impacts—for example, energy-saving lights installed (and left on) in places there were no lights before, such as a garden or garage, or the unintended effects of symbolic consumption, such as the discarding of perfectly functioning goods when their social appeal ends. Current approaches to eco-design, which focus on physical properties of things at the expense of their psychological impacts, do not really help us make sense of these unintended consequences. The solution, according to Zhang, is to design not only sustainable products but also sustainable behaviors—a "both/and approach" to eco-design. The goal of the both/and approach and of eco-design is not just to create green products but also to design sustainable behaviors, particularly lasting attachments that can overcome the pull of symbolic consumption and waste. Successful eco-design creates relations between humans and products for the sake of the environment.

Marc Sagoff and Philip Brey continue the discussion of product design by carefully examining the relationship among production, consumption, and the natural world. They resist the temptation to decry consumerism as resource depleting, wasteful, and polluting, if not empty and soul-sucking. Instead, they attempt to understand consumerism as a union of technological, social, and environmental issues. They find no causal relationships or simple patterns among them, but instead find complex associations of moral, political, and economic forces. Consequently, any proposed environmental changes in production or consumption cannot be simply designed away; they have to take into consideration all the complex ways that people make, use, and dispose of things.

Sagoff examines the relation between sustainability and the production and consumption of consumer products. He takes the optimistic view that economic production will never be seriously constrained by a lack of natural resources. When a resource becomes scarce, its price increases, leading us to substitute plentiful resource flows for scarce ones, to do more with less, and generally to continue to improve standards of living by advancing technology. That is, technological advances will ensure that natural capital will always be cheap and fungible. None of the concerns that have occupied the environmental movement since the 1970s—global population, depletion of non-renewable resources, or food shortages—have materialized. Sagoff suggests that environmentalists embrace technological solutions instead of denying the power of technological progress or simply decrying consumerism as wasteful. Nevertheless, he suggests that there are indeed good reasons to question consumerism. Although technology can overcome the physical limits nature sets on the amount we can produce and consume, there are moral, spiritual, and cultural limits to growth. Simply put, we consume too much—not because of the resources we use but because our market-driven consumerist culture undermines "the bonds of community, compassion, culture, and place." We consume too much when consumption becomes an end in itself and "makes us lose affection and reverence for the natural world." Sagoff wishes to focus the debate on consumerism on the social lives we seek to preserve rather than the resources we may exhaust. That way we might stop vilifying technology and romanticizing nature.

Brey also examines the relationship between sustainable development and consumerism, but he focuses on the role that technologies play. He begins by discussing ecological modernization, the current development practice that aims at greening production and the global economy in ways that leave existing economic and political institutions intact as much as is possible. He agrees with critics who claim that sustainable development is incompatible with modernization's ideal of unlimited growth. A more fundamental reform of development must also transform patterns of consumption and challenge the values and beliefs that underlie consumerism and materialism. The development of sustainable consumer products should, therefore, promote sustainable behaviors and lifestyles, as well as reduce or eliminate consumer products that are unsustainable and which promote unsustainable behaviors and lifestyles.

In addition to these ecologically designed green products, there are products designed to change the attitudes or behaviors of users through persuasion and social influence. Persuasive technologies, such as showers that turn off after five minutes of use and cars that encourage economical use of fuel, requires certain actions and precludes other actions. But Brey worries that the redesign of technologies to promote sustainable consumption, though necessary, will not be sufficient to engender sustainable systems of consumption. The idea that this is possible amounts to another belief in a technological fix, this time by the social engineering of lifestyles and patterns of consumption through a reform of technology. Technological reform will certainly be of great help in the move toward sustainable patterns of consumption, but it should be seen as part of a comprehensive strategy that also includes social and economic incentives and public debates about values, lifestyles, and the future of the planet.

The final two chapters are concerned with agricultural ethics, a field that is not usually considered to be a part of the philosophical literatures on technology and the environment. Agricultural ethics deals with the normative dimensions of farming, food processing, resource management, and distribution. As with the issues of design, sustainability, and consumption, agriculture is inseparable from the technologies it relies on and the environments in which it functions. There is no food production without devices or machines, nor is there food production without effects on ecosystems and animals. Furthermore, agriculture produces consumer goods, and thus it confronts issues of sustainability and design similar to those that affect non-edible products. The chapter by Raymond Anthony and that by Wyatt Galusky address animal agriculture: the industrialized production of meat, milk, and eggs for human consumption. The main ethical considerations are animal welfare, environmental sustainability, and the role of consumption—in other words, how the design and the governance of technologies affect the health and welfare of humans, animals, and the environment.

Anthony addresses the role of technology in promoting sustainable agri-food systems and animal well-being, relating the works of philosophers concerned with the excessive commodification of life to the common view that farmed animals are mere resources. Food production under an industrial model, fueled by market imperatives to expand and reduce costs, makes it difficult to appreciate the good of animals in their own right, apart from their instrumental use for us. They are seen as mere commodities, like any commercial object. The problem is not merely our attitude toward animals but the very modes of production in which we deal with them. Agricultural technologies reflect our values and norms, for better or worse.

Anthony suggests a virtue-ethics approach to technology in order to counter the instrumentalist view we typically have about man-made things. If virtues can be embedded within machinery, it may be possible to design agricultural systems that can recognize the intrinsic good of animals. An environmental virtue ethics of care (EVEC) is the antidote to commodification of humans, animals, and the natural world. According to Anthony, EVEC not only affirms ethical consumerism but also requires that industry technocrats act virtuously in their relationships with animal agriculture. They ought to be mindful of how they innovate, what products they market, how they design facilities, and, above all, how they might find better ways to meld business, profit, and technology with care for humans, animals, and the environment.

Galusky examines the role of technology in producing meat for human consumption, rehearsing arguments against industrialized animal agriculture and arguments in defense of *in vitro* (laboratory-produced) meat. But Galusky complicates the idea that technology solves the problems of factory farming by considering meat *as* a technology, not just a product of it. He does this in order to understand meat as a human creation that involves a network of relationships among technologies, ourselves, and the natural world—that is, a certain kind of process that we are involved in and therefore are responsible for. When we view meat as technology, we highlight the worldviews, contexts, and agents that make it possible. For Galusky, these include the view of the natural world as "plastic," the ultimate virtue of control over nature, and the view of animals as merely sources of protein. Industrialized meat technologies raise ethical question about what kind of nature, what kind of humans, and what kind of animals we are designing. He reminds us that the more technologies we make, the more responsibilities we take on.

It is my hope that readers will take note of both the similarities and the differences among these chapters. They should find a broadly shared focus and scope of concern about both technology and the environment, as well as a genuine desire to transcend narrow academic subdisciplines to address common concerns. But there are also important questions to be asked and significant contrasts to consider:

- Is Callicott right when he suggests we should stop trying to identify underlying conceptual presuppositions about the environment, or is Ihde right when he says we do not yet understand enough about techno-environmental problems given their ambiguities and complexities?
- Who is right about techno-pessimism and fears: Callicott and Ihde for finding them unwarranted, or Whyte, Gunderson, and Clark and Thompson for finding them occasionally quiet well warranted?
- Is Thompson correct that the wrong kind of scientific risk assessment—and lack of engagement with social justice issues—is responsible for our irrational techno-fears?
- What about Hale's deontological justification for morally legitimate environmental remediations and the Oxford Principles? Are such remediations and the Oxford Principles compatible and complimentary, or do they represent different models for geoengineering research and development?
- What does Allenby's skepticism about industrial ecology mean for van de Poel's attempt to identify design requirements for sustainable technologies?
- Is Wei too optimistic about how eco-design can design sustainable behaviors? Is Sagoff too optimistic that technological improvements will always overcome depletion of natural resources?
- Does Brey put all the discussions of sustainable design and consumerism into perspective by showing how they are at best status-quo affirmations of ecological modernization and at worse oversimplifying techno-fixes to entrenched socioeconomic systems?
- If Brey succeeds in doing so, does that have bearing on Anthony's call for an ethics of care and Galusky's plea for greater responsibility for industrialized animal agriculture by producers and consumers? Are ethical solutions adequate for political problems?

In addition to questioning each chapter, I leave it to the reader to determine whether this volume as a whole has achieved its aims of examining how technology is central to environmental philosophy and how the environment is central to philosophy of technology; how philosophical discourses on technology and the environment are intimately related; and how it is possible to create a hybrid discourse that might foster dialogues among other academic disciplines and among people other than teachers and students. It is my hope that the volume contributes not only to the philosophical conversation about how the philosophy of technology relates to environmental philosophy but also to broader conversations about how technology can and should relate to the environment.

Note

1. Other journals have published some cross-specialization articles but not nearly as many. The other major journals include *Philosophy & Technology* (Springer) and *Science, Technology, and Human Values* (Sage); *Environmental Philosophy* (Philosophy Documentation Center), *Environmental Values* (White Horse Press), and *Ethics, Policy & the Environment* (Taylor and Francis).

References

Banchetti, M., Lester Embree, and Don Marietta, eds. 1999. *Philosophies of Technology and the Environment.* JAI Press.

Miller, Glen. 2015. Mapping Overlapping Constellations: Nature and Technology in Research in Philosophy/Techné and Environmental Ethics. Doctoral dissertation.

White, L., Jr. 1962. *Medieval Technology and Social Change.* Oxford University Press.

1 Back to the Future: The Return of STS to Its "Historical Roots"

J. Baird Callicott

It has become increasingly clear to everyone in the field that "The Historical Roots of Our Ecologic Crisis," by Lynn White Jr., is the text that spawned academic environmental philosophy, for better or worse—and some think it was for the worse (Norton 2003). That article was published in *Science* in 1967, but it took several decades for the appreciation of its importance for the field to sink in. That's because the most salient impression left by White's screed is its lurid and cavalier claim that Christianity "bears a huge burden of guilt" for the environmental crisis (White 1967). How could that be? Because, according to White, the Bible, in Genesis 1: 26–28, creates man alone in the image of God, gives him dominion over the rest of the creation, and commands him to subdue the Earth. Living out this collective fantasy, the civilization that evolved in a climate of thought dominated by Christianity produced all the technological wonders we presently enjoy, but also all the environmental insults that now threaten to destroy that civilization, indeed civilization in general. Of course, the doctrines at issue are Judaic not peculiarly Christian. But because Christianity adopted the Hebrew Bible as its "Old Testament," for centuries Christians also adopted the belief, White insisted, that man had a God-given right ruthlessly to exploit nature for his own ends. Thus, most of us environmental philosophers thought that the influence of "Roots" was limited to a fairly narrow debate about how Jews and Christians past read these verses (and others) of the Bible and, more to the point, how Jews and Christians present might read them now and in future. Should they be read as an invitation to a despotic tyranny over the Earth, as White proposed, or as a call to be stewards of God's Creation, as Christian and Jewish apologists proposed.

Beneath White's notorious thesis there lies a persistent subtext—a refrain repeated, variously formulated, five times in the article:

• What shall we do? No one yet knows. Unless we *think about fundamentals*, our specific measures may produce new backlashes more serious than those they are designed to remedy. As a beginning we should try to *clarify our thinking* by looking, in some historical depth, at *the presuppositions that underlie modern technology and science*. ... (p. 1204)
• The issue is whether a democratized world can survive its own implications. *Presumably we cannot unless we rethink our axioms*. (p. 1204)
• *What people do* about their ecology *depends on what they think* about themselves in relation to things around them. Human ecology is deeply conditioned by beliefs about our nature and destiny—that is, by religion. This is very evident in India and Ceylon. It is equally true of ourselves and of our Medieval ancestors. (p. 1205)
• *What we do* about ecology *depends on our ideas* of the man-nature relationship. More science and technology are not going to get us out of the present ecologic crisis until we find a new religion or *rethink* our old one. (p. 1206)
• *We must rethink* and refeel *our nature and destiny*. (p. 1207]

White, of course, abuses the word *ecology* here. Most people don't do anything about their ecology. Because most people are not ecologists, they do not have an ecology to do anything about. White might more precisely have asked "What people do *in respect to their environments* depends on what they think about themselves in relation to things around them?" and "What we do *in respect to our environments* depends on our ideas of the man-nature relationship?" He also distracts attention from his point by conflating "beliefs about our nature and destiny" with "religion." Not all beliefs about our human nature and human destiny are religious. And religion involves much more than beliefs about human nature and human destiny. In any case, to meet the challenge of the environmental crisis more science and more technology, by themselves, are not going to do the trick; indeed, more science and technology might just make matters worse, White declared. First and foremost, we must "rethink our axioms." That's because "what we do" in and to the natural environment depends on "what we think"—what we think about nature, human nature, and the relationship between humans and nature.

Almost subliminally—because this more general subtext was obscured by his apparent excoriation of the allegedly "orthodox Christian view"—White thus set out an implicit agenda for a future revolutionary environmental philosophy. We had approached nature confident in our belief that it was in essence a Newtonian machine. We believed ourselves to be junior-grade designers and artificers—created in the image of the Big Designer/Artificer in the sky—licensed to redesign nature to suit ourselves. The environmental crisis is, as it were, nature's mute way of talking back. Thus we have to *rethink the nature of nature, human nature, and the relationship between humans and nature* as the first and most important step toward meeting the challenge of the environmental crisis. That's what White was whispering in the ears of a few young philosophers who had come of age in the Age of Environmental Crisis.

After all, whose professional remit is it to rethink conceptual presuppositions, fundamentals, and axioms? Philosophers, of course! Saving the world from the environmental crisis—I distinctly remember it dawning on me back in 1969 or 1970, when I first read "Historical Roots"—was *my* responsibility, because I was a philosopher and rethinking (that is, critically engaging) presuppositions, fundamentals, and axioms is what philosophers, in particular, do. It's worth repeating: To the end of saving the world, the principal task, our task, White seemed to tell us, as nascent environmental philosophers, is to rethink fundamental, axiomatic presuppositions about the nature of nature, human nature, and the relationship between humans and nature. Then, we environmental philosophers could inform engineers, who, guided by us, could begin to design "appropriate technologies," as they soon came to be called. The early 1970s was a heady time for the first generation of environmental philosophers. White had given us not only an agenda, but a mission. Saving the world from ecological doom was up to us—or so it then seemed.

This implicit agenda had two phases, the first critical and the second reconstructive.

White himself had begun, but had only begun, the critical phase by critiquing the notions of nature, human nature, and the relationship between the two in what he claimed was the Judeo-Christian heritage of ideas. But that is not our only heritage of ideas. What about the Greco-Roman heritage, classical atomism, Pythagorean/Platonic otherworldliness and mathematical formalism, the Aristotelian/Thomistic hierarchical means-ends structure of nature, Cartesian dualism, Lockean private property, Kantian moral rationalism, utilitarianism, Berkeleyan idealism, Logical Positivism, etc., etc.? The pages of the early volumes of the journal *Environmental Ethics* are filled with indignant diatribes focused on various philosophical systems in the intellectual tradition that begins with the ancient Greeks.

How to go about the second phase was hinted at by White himself, but not as well articulated by him as how to go about the first phase. With what can we replace the erroneous ideas we have inherited from both our Judeo-Christian and our Greco-Roman intellectual roots about nature, human nature, and the relationship between humans and nature? White suggested two alternatives: to mine the conceptual resources of non-Western traditions of thought, such as Zen Buddhism, and to recover minority strands of ideas in the Western tradition that he compared, in a subsequent article, to "recessive genes" (White 1973). We might call them recessive memes, to borrow a later coinage by Richard Dawkins. White himself identified the Franciscan doctrine of the animal soul as a promising recessive meme in Christianity. Improbably, Arne Naess found environmentally sound recessive memes in the philosophy of Spinoza. Heidegger's *Being and Time* was touted by Michael Zimmerman as a repository of notions that might save the environment—until Heidegger was outed as a Nazi, after which Zimmerman recanted. And so the pages of the early volumes of *Environmental Ethics* were filled with debates about the potential of Daoism, Hinduism, Pythagoreanism, and so on and so forth—as well as of Zen Buddhism, Spinozism, and Heideggerianism—to save the environment, if only we would all become converts to this exotic worldview or that recessive system of memes.

Lynn White Jr. was not a philosopher by trade. Rather, he was a historian of technology, specializing in the late Medieval period. Much of "Roots" is devoted to bits of arcana in that arcane field. For example, he devotes a great deal of attention to the effect on the environment of the eight-oxen-powered, deep-digging, earth-turning, mold-board plow that sliced, diced, and laid bare the rich, heavy soils of northern Europe in the late Middle Ages. In passing he mentions harnessing flowing water and blowing wind—to the ultimate end of saving human and animal labor—and "that most monumental achievement in the history of automation: the weight-driven mechanical clock" (White 1967, 1204).

It is something of a leap to claim that several verses of the Bible caused the environmental crisis. And while this is what most readers take away from "Roots," White actually traces out a causal chain linking the two. And in that putative causal chain we might find an agenda for a future integration of environmental philosophy with the philosophy of technology.

The proximate cause of the environmental crisis is "*modern* technology," White claimed, soberly noting that human technology is age old and, indeed, even that *all organisms* have environmental impacts (some, such as beavers, more than others). But modern technology is so much greater in the degree of its environmental impact that it has become different in kind from all past vernacular technologies. Is that true?

What makes modern technology modern, according to White, is that it is informed by science. Pre-modern vernacular technologies, he claims, were purely empirical, developing by trial and error, among yeoman artisans who worked with their hands. Pre-modern science (that is, natural philosophy/theology), by contrast, was purely speculative and theoretical, the pursuit of knowledge for knowledge's sake, by leisured aristocrats, who disdained manual labor (and laborers). First the Reformation and later the rise of an urbane capitalist middle class and finally the replacement of feudal aristocracies by democracies permitted the "marriage" of science and technology. Is that true? If it is, the experimental method, the quintessential difference between modern and pre-modern science, depends on that marriage of science and technology—although White himself doesn't mention it. Experiment requires instruments and apparatuses and instruments and apparatuses are a form of technology. The "marriage" of science and technology—of "head" and "hand"—was, if indeed it occurred as White's account purports, one of mutual convenience. The twins Modern Technology and Modern Science were its issue.

White goes on to claim that both science and aggressive, powerful technologies—such as that eight-oxen-powered, deep-digging, earth-turning, mold-board plow and water-powered mills and other hydrolic machines—are of Western provenance, indeed more specifically of Western Roman Catholic European provenance (as opposed to Eastern Greek Orthodox European provenance). The implication is that not only the Byzantine civilization, but the civilizations of India, China, and the rest of the Asian civilizations did not also give birth to science as we know it or to aggressive, powerful technologies. Is that true?

Because both science and technology developed exclusively in Christendom—if indeed they did—the Christian worldview must have nurtured them, according to White. How so? Here's where we get to the link between science, technology, and verses 26–28 in Genesis, chapter 1. God's first, unnumbered commandment—to have dominion over and to subdue the Earth—motivated the development of aggressive and powerful technologies. And because we are created in the image of God—which image began to be theologically associated with (Greek) reason in the late Middle Ages—we could reverse engineer the creation and recreate the rational blueprint that God had in mind as he created the world. The "natural theologians," of the seventeenth century—among them Issac Newton—said as much. They essayed to "think God's thoughts after Him." Is all that true?

The deeper question for the philosophy of technology is White's subtext that what we do depends on what we think. It was challenged by the geographer Yi-Fu Tuan (1968). But most of the first generation of environmental philosophers ignored Tuan's counter examples or sought to demean the skeptical conclusion that Tuan drew from them. By now, of course, that deeper question should be raised for critical reflection: Is it true that what we do in and to the natural environment depends on what we think about nature, human nature, and the relationship between humans and nature? Environmental philosophers readily accepted that proposition; or, rather, it would be more accurate to say that we ourselves went about our business without raising that proposition up for sustained reflection, let alone critical examination. We felt, albeit subliminally, that Lynn White Jr. had anointed us to be the most preeminent of all would-be world savers. For if the only way we can meet the challenge of the environmental crisis is by rethinking our axioms, that was our job to do? One way to put this question is "Should we be disabused of our megalomaniacal pretenses?"

There is a counter-proposition: technological determinism. Maybe what we think depends on what we do instead of the other way around. Or more precisely put, maybe what we think depends on the technologies with which we do what we do. Ironically, White himself was a proponent of some sort of technological determinism in his 1962 book *Medieval Technology and Social Change*. In that book he propounds the "Stirrup Thesis," summarized in these words: "Few inventions have been so simple as the stirrup, but few have had so catalytic an influence on history. The requirements of the new mode of warfare which it made possible found expression in a new form of Western European society dominated by an aristocracy of warriors endowed with land so that they might fight in a new and highly specialized way." The stirrup thesis is analogous in the history of technology to Frederick Jackson Turner's "frontier thesis" in the history of the American west. It precipitated a paradigm shift; it was, for a time at least, a game changer.

Perhaps such technologies as the stirrup determine such things as social orders or classes, but do they determine what we think? That is the question. If we are inclined to believe that White's *technological determinism* in *Medieval Technology and Social Change* is correct, we might also be inclined to think that the purely empirical—that is, uniformed by aristocratic science—yeoman mechanical inventions in the thirteenth century, as they gradually spread in subsequent centuries, inspired the articulation of a mechanistic worldview by Descartes and Newton in the seventeenth century. These natural philosophers were just giving metaphysical expression to the new technological *esprit* that informed them, rather than the other way around. On the other hand, if we are inclined to buy White's *intellectual determinism* in "Historical Roots," we might suppose that the mechanistic worldview of Descartes and Newton was, as he explained, the product of Christian theology—belief in a divine Creator who created man in His own (rational) image—informed by Greek natural philosophy, more especially

Democritean atomism and Pythagorean-Platonic mathematical formalism. From this point of view, it was only after the articulation of a mechanistic metaphysics in the seventeenth century that a thoroughly mechanized technology (and society) emerged, beginning with the Industrial Revolution in the eighteenth century and reaching its zenith in the twentieth century.

I am inclined to think that a positive feedback loop—a kind of mutually reinforcing historical dialectic—goes on between what we think and what we do, that is, between natural philosophy and technology. In the history of Western civilization it starts, I believe, with what was thought, Democritean atomism and Pythagorean/Platonic formalism in ancient, aristocratic natural philosophy. With some inspiration derived from the invention of mechanical technologies in the late Middle Ages, a robust mechanistic metaphysics was articulated in the seventeenth century by Descartes and Newton, driven mainly by purely intellectual forces—the recovery of Greek natural philosophy (in reverse order, first Aristotle, then Plato), the discovery of the Arabic developments of it in the late Middle Ages, and then the Copernican revolution in the sixteenth century. The mechanical philosophy, in turn, inspired and informed the explosive development of mechanical technologies in the eighteenth century, which reinforced the mechanistic worldview in the nineteenth and twentieth centuries.

In the mid to late years of the twentieth century, solid-state electronic technologies (televisions, computers, cell phones) emerged alongside mechanical technologies. This development is consistent with my own inclination toward a think-first-do-after-reinforcement-of-what-was-first-thought historicism. These technologies—most indisputably television—were technological applications of the second scientific revolution in the late nineteenth and early twentieth centuries. Therefore, we first have a paradigm shift in natural philosophy followed by a paradigm shift in technology. Now, solid-state electronic technologies may be reinforcing the more holistic electromagnetic field and quantum-theoretical metaphysics implicit in the second scientific revolution.

The most radical technological determinists—Walter Ong, Eric Havelock, and Marshall McLuhan—claim that communications technologies don't merely transform this or that idea, or even this or that system of ideas; they profoundly transform human consciousness itself (Ong 1982; Havelock 1986; McLuhan 1962). According to Ong and Havelock, the invention of alphabetic writing made possible the abstract idea, encouraged reductive habits of mind (as in fifth-century atomism, which Aristotle, in the *Metaphysics*, tellingly illustrates with letters of the Greek alphabet), created individuality, and, with the emergence of individuality, a dread of personal death (as evidenced by the preoccupation of the newly literate breed of abstract, individualistic thinkers, such as Pythagoras and Plato, with speculations on the immortality of the soul and by their cogitation of elaborate eschatologies). Human consciousness in a state of orality, by contrast, involved mythopoeic narrative thought, holistic habits of mind, communal identity, and, accordingly, indifference to personal mortality. According to McLuhan, the invention of the printing press only served to intensify the revolution in consciousness created by literacy.

Ong and especially McLuhan speculated on the revolution in consciousness that twentieth and twenty-first-century communication technologies might engender. The two were born only a year apart (1912 and 1911, respectively), and the first post-print communications technology to which they were exposed was radio. The prevalence of radio during their formative years suggested a shift from a visual to an aural communications orientation. Which further suggested a shift from an analytic, reductive, and rational consciousness to the more communal, mythopoeic, and irrational consciousness of orality. Radio, as it were, was creating a neo-oral human consciousness. But radio, although still alive and well, has been succeeded by television and a multitude of online electronic communications technologies. The Ong-Havelock-McLuhan brand of radical technological determinism may be correct, but predicting the nature of the revolution in human consciousness that post-print technologies of communication will effect is, I am inclined to think, totally speculative.

If Ong, Havelock, and McLuhan are indeed correct, it will not be the likes of Arne Naess, Holmes Rolston, Paul Taylor, and Val Plumwood who can provide us with the wherewithal to meet the challenge of the environmental crisis. It will, rather, be the likes of Guglielmo Marconi, Philo Farnsworth, Steve Jobs, and Bill Gates. The twentieth century gave birth to new communications technologies (radio, television, the cell phone, the Internet), and the twenty-first century is giving birth to more (the smart phone, Facebook, cloud computing, Twitter). If these technological determinists are right and these communications technologies will have as profound an effect on human consciousness as they claim that alphabetic writing and print had in the ancient and modern periods, then our ideas of nature, human nature, and the relationship between them will be revolutionized—whether traditionalists like it or not. In that case, we environmental philosophers will just have turned out to be the cheerleaders—not the world-savers or our megalomaniacal fantasies—narcissistically nattering on the sidelines of a technology-driven ecology of mind (if that's what it turns out to be).

References

Havelock, Eric. 1986. *The Muse Learns to Write: Reflections of Orality and Literacy*. Yale University Press.

McLuhan, Marshall. 1962. *The Gutenberg Galaxy: The Making of Typographic Man*. University of Toronto Press.

Norton, Bryan G. 2003. *Searching for Sustainability: Interdisciplinary Essays in the Philosophy of Conservation Biology*. Cambridge University Press.

Ong, Walter J. 1982. *Orality and Literacy: The Technologizing of the Word*. Methuen.

Tuan, Y-Fu. 1968. "Discrepancies Between Environmental Attitude and Behavior: Examples from Europe and China." *Canadian Geographer* 12: 176–191.

White, Lynn, Jr. 1962. *Medieval Technology and Social Change*. Clarendon.

White, Lynn, Jr. 1967. The Historical Roots of Our Ecologic Crisis. *Science* 155:1203–1207.

White, Lynn, Jr. 1973. Continuing the Conversation. In *Western Man and Environmental Ethics*, ed. Ian Barbour. Addison-Wesley.

2 Phil-Tech Meets Eco-Phil

Don Ihde

Philosophy of technology (phil-tech), relatively new to the North American philosophical scene in the seventies, emerged from its largely European roots under a somewhat dark cloud of technophobic colors. The godfathers, Martin Heidegger, Jacques Ellul, and Herbert Marcuse as the most popular, portrayed technologies as Technology, a sort of transcendental dimension that posed a threat toward culture, created alienation, and even threatened a presumed essence of the human. Although this dystopic tendency was later moderated by younger American philosophers of technology who saw multiple possibilities for admittedly nonneutral technologies, the more pragmatic and more empirical effect took some time to mature.

Roughly concurrent, but somewhat later to arrive as a parallel and related subdiscipline, philosophy of the environment (eco-phil) began with what I take as a similar dystopic perspective. Early alternative technology worries over neo-Malthusian population explosions and unsustainable consumer practices were often at the core of the eco-philosophies.

I wish to undertake a pairing of phil-tech and eco-phil with respect to the worries often dominating these reflections on contemporary life. I wish to look at some more nuanced situations regarding utopic and dystopic tendencies, and suggest directions that might be developed in relation to the hopes that both fields might entertain.

Although I was not quite a charter member of the Society for Philosophy and Technology (SPT), I have been active in this group since the late seventies. And to set the tone for phil-tech, I want to recall my first entry into this group. I had been asked to participate on a panel responding to the work of Hans Jonas, whose work concerned the early days of biotechnology and related genetic and sometimes environmental studies. At first I did not know what issue to address until I read the passages that related to Jonas's call for a new ethics, an ethics of fear as he called it, as the appropriate response to the biological manipulations that he perceived as threatening the very notion of a human essence. My role, then, became that of critic with an attack upon yet another Euro-American example of technological dystopianism.

I cite this example to illustrate what I take to be a deep set of intellectual habits that seem to be common to many both in environmental studies and in much philosophy of technology: congenital dystopianism. In the meditations to follow, I am going to look at a series of these intellectual habits that are commonly held between environmentalists, or ecologists, and philosophers of technology, and in each case give certain critical responses with the aim of redirecting concerns that should unite these two subdisciplinary matrices.

The Rhetoric of Alarm

Within the precincts of SPT, the best-known institutional group for philosophy of technology in North America, many commentators have noted the dominance of the dystopian. If there are godfathers of SPT, they have been Ellul, Heidegger, and the Karl Marx of industrial capitalistic alienation. Every one of these godfathers, at least as interpreted within the SPT context, displays some variant upon the ways in which Technology has become the degrading metaphysics of late modernity and, insofar as environmental issues enter the scene, is taken to be, in industrial embodiments, the primary cause of environmental degradation.

But the same themes often dominate environmental precincts as well. In Lester Brown's introductory chapter to *The State of the World*, titled "The Acceleration of History," the now-familiar refrain, borrowed and extrapolated from Thomas Malthus, is rehearsed: There have been more humans on Earth in the last 30 years than in all previous human history (overpopulation), there is a dwindling food supply both in oceanic fishing resources and in agriculture (Malthusian arithmetic versus geometric extrapolations), and concerns for global pollution accelerate the degradation process (Brown 1996).

I am terming this the rhetoric of alarm. It is correlated to Jonas's ethics of fear and its purpose is—in Paul Revere fashion—to awaken the listener to the dire fate of a presumed environmental catastrophe, the late modern equivalent of the redcoats.

Historically, the rhetoric of alarm is the flip side of nineteenth-century progressivist utopianism. It seems hard to believe, but there were accounts in England that extolled the more brilliant sunsets of the Industrial Revolution, already known then to be caused by industrial haze in the atmosphere! Indeed, it may well have been that the utopian promises of industrialization and technologization at the turn of the twentieth century, by their very over-extrapolation, led to part of the flip phenomenon in the middle of the twentieth century. The promises of technological solutions to social problems, surpluses of food through agricultural revolution, democracy for everyone through communications, none of which occurred in either the time predicted or to the degree expected, may well have helped the distracters who linked technologies to everything from warfare to the Holocaust.

How, then, do we escape the horns of the utopian/dystopian dilemma?

Response: Critical and Skeptical Cautions

I am, of course, suggesting that excessive rhetorical strategies may work both ways. If promising too much can lead to disillusionment, then prophesying results that do not occur may lead to apathetic responses. Here my diagnostic will combine what I take to be the best habits found within scientific communities with those of a critical philosophy.

Excessive rhetorical strategies often are ill-founded and cause more harm than good. My first example for critical demythologization is the notion of a Malthusian extrapolation. In its original form, it was a hypothesis that populations would, in some time period, exceed the capacity of the Earth to feed them. In the long term, this could not occur if for no other reason than that excessive populations would cause starvation until a lower population level would return to some kind of homeostasis. Malthus himself eventually recognized this and modified his earlier theses.

Moreover, the latter-day version of such an extrapolation is even more ridiculous than its original Malthusian form. The fact of the matter is that before the eighteenth century there was no way of accurately estimating the size of the human population. This is not merely because we did not have census statistics for prehistory, but because I contend there is often a tendency to undercount, by virtue of another set of bad habits relating to historical foreshortening even within science, which oversimplifies prehistorical trends. Did you know, for example, that the pre-colonial populations of the Americas are currently estimated to be hundreds of times larger than the estimates of only 30 years ago?

On a visit to Mexico, I got a look at the world's largest pyramid, located in Cholola. Its base is more than twice that of Cheops, but this pyramid was not even known until 1910, because it is buried under thousands of tons of dirt deliberately placed upon it. Surmising that if there is one such buried pyramid there might be more, I entered into a conversation with a Mexican anthropologist and asked him if he knew of other buried pyramids. His response was that recent air surveys (using magnetometers that can detect undersurface structures) show that there may be 86,000 such pyramids in northern Mexico. In short, the probable population of northern Mexico in pre-colonial times just took a more than Malthusian leap in size. A similar population leap took place with the discovery in the last decade by similar processes of many cities of up to 10,000 people in the Mississippi Valley of the last millennium, one of which existed in Cahokia, Illinois. This city was larger than Copenhagen at the same time a millennium ago.

Another example relates to the so-called Neanderthals. Although recent discoveries show that there was a very widespread distribution of these near relatives of modern humans, everything we know about them is based upon only about thirty skulls, widely distributed. We simply do not know what population numbers to attach to this hominid species.

J. Robert Oppenheimer claimed, in the 1950s, that there were more physicists alive then than in all pre-twentieth-century physics, to which Art Buchwald replied "At that rate there will soon be two physicists for every person alive!" In short, a dose of good old-fashioned philosophical skepticism concerning historical accelerations may be warranted. The same errors of scale occur on the future side of the extrapolation. As of 1997, the number of nations fallen below self-replication in population had risen alarmingly in the previous decade. While we have known for some time that industrialization is often followed by a population decrease—the average number of children per family in both the United States and Japan has shrunk from more than 5 before World War II to 2.2 or fewer since—the lowering of growth into negative numbers is still more recent. There is now evidence that similar falling birthrates are occurring in South American nations, most of which have fallen into a range of 2.1 to 2.8 children per family, not far from the 2.2 figure needed for a level population. Thus, were we to apply the negative reversal projection in Malthus/Oppenheimer/Buchwald fashion, might we not be justified in worrying about the self-reduction of the human population to zero by the end of the next millennium? Moral of the story: Remain skeptical of extrapolative arguments that may underestimate pasts and overestimate futures.

Microsolutions to Macroproblems

An intellectual habit found in both philosophy of technology and ecological circles of philosophers, which has been applied, in my opinion, as badly as excessive rhetorical strategies, is the tendency to see problems as macroproblems but to propose microsolutions. Perhaps this is a variant upon "think globally, act locally."

Waste and recycling is an example of macroproblem/microsolution thinking. Waste and chemically toxic processes are clearly macroproblems, particularly acute in industrialized nations. Toxic waste (and toxic applications to agriculture, etc.) are the most extreme cases of this problem. What are the solutions? The most popular microsolution is recycling of waste, and this is accompanied by the usually ineffective solution of straightforward banning of toxic products.

I am not against recycling or bans. I am merely indicating that the microsolution by itself does not solve the macroproblem. Even if recycling were to rise in efficiency from its current 15–20 percent effectiveness to some future 60–70 percent effectiveness, it would merely slow, not stop, the problem. Similarly, the banning of all toxic materials without substitute would not solve the problems of either cleaning up the environment or raising food productivity. It would, instead, displace the problems into different contexts since the bugs, microorganisms, and diseases that had previously been partially controlled would reassert themselves.

What I am pointing to is the tendency, the intellectual habit, to think "small is beautiful," which is, to my mind, equivalent to a form of nostalgic romanticism found among philosophers of technology and ecologists. For example, in the 1970s the preferred solution to the globalization of technologies among the dominant groups of philosophers was to favor "appropriate technologies," i.e., small and simple technologies for (ignorant, untrained, and unready) Third World peoples. Given both the implicit romanticism and condescension involved, it is little wonder that such policy recommendations failed.

Today, that solution is currently replaced with the search for sustainability, a systems approach that seeks stasis (or, as one of my Dutch colleagues termed it, a new search for perpetual motion.) My experience of responses, when conferences on this issue are held, indicates that success in sustainability is as unlikely as was appropriate technology 20 years ago. The reasons are not hard to discern, since so long as appropriateness and sustainability apply unequally to the First and Third Worlds, there is likely to be little appreciation for the microsolution. As long as whole populations remain reluctant to deliberately lower living standards, the equalization process between the First and Third Worlds will not take place.

Another, although more minoritarian, version of a microsolution for a macroproblem, again found within both philosophy of technology and environmental circles, is a form of retrospective romanticism. Here the shape taken is one that idealizes the microsolutions of indigenous peoples or of traditional cultures. David Abram, won a major literary prize, the Lannon Literary Award, for his book *The Spell of the Sensuous*. Abram draws from his experience of traditional cultures—mostly Native American but also many smaller Asian cultures—to argue that forms of respect for the environment and nondestructive practices still thrive among such peoples (Abram 1996).

There are two problems with this retrospective romanticism. The first is that even where the microsolutions did work, they worked in contexts that largely no longer exist. Slash-and-burn agriculture can indeed work if and only if the jungle area is large enough to recover between migrations of the agriculturists. But to maintain this situation, both territory and population size must be maintained. Populations that have increased, whether globally or locally, by virtue of the medical technologies that lower infant death rates and increase the number of adults who live into old age (never mind utopian wish fulfillment urges to longer lives), soon change this situation. Few romanticists actually advocate giving up all modern medicine! Nor do I find much appreciation for the massive bloodletting rituals of war and slavery that may have contributed to keeping overpopulation from occurring in Mesoamerica.

The second problem is that while some cultures did indeed attain human and environmental sustainability for very long periods of time (Australian Aboriginals, Arctic Inuit, European peasants), one should not overlook the many more societies whose premodern practices ended up in human-initiated environmental degradation (Mesopotamian irrigators, Easter Islanders, and, earlier, the humans who may have contributed to the extinction of the large mammals of the late Ice Age, or, more recently, our own civilizational ancestors who deforested the entire Mediterranean Basin). Not just any traditional culture or indigenous people can attain sustainability.

Response: Critical and Sketpical Considerations

What I am really arguing is that we have not yet fully diagnosed either what our technologies can or should do, or what the environmental crises are. So long as we naively accept both negative and positive hype that there will soon be 10 billion humans on Earth, or that we will soon be a chummy global village through the Internet, we are not likely to bring either technologies or ecosystems into appropriate focus.

Admittedly, to this point I have exercised my own style of acerbic, somewhat cynical critique. Perhaps the time has come to turn to the positive side of the analysis and make some suggestions for how philosophers of technology and ecologists can begin to address the environmental problems we face.

Sizing Up the Problems

I suggested earlier that I would draw upon aspects of the scientific community at its best and its subsequent impact upon policy. Regarding problems of the environment, I first wish to point to the process and result of scientific consensus building. In today's global scientific communities, there are several recent consensuses. The planet is warming up as a whole, and there are now well-established signs of a greenhouse effect. This consensus, however, has been hotly debated and contested, particularly with respect to the precise degree to which homogenic causes contribute. But with the hyper-high-tech measurements we are now able to make, the scientific community has at least demonstrated that global temperature has risen, that ocean levels have risen, and that global weather patterns have changed in keeping with models of a greenhouse effect.

Within this consensus, hardly news to most of us today, one particularly recent and challenging set of observations relates to what could be called a middle-sized problem: the ozone hole. This phenomenon is now known to be exacerbated by the rise in chlorofluorocarbon gases (CFCs), carbon monoxide, etc., many of which can be traced to particular industrial products.

This middle-sized problem, however, did not fall into the doomsday dystopian predictions, nor has it necessarily disappeared. Instead, it was recognized, publicized, discussed by the governments of the world, and with a relatively short political life span, got addressed through global legislation and agreements that appear to have already begun to slow or eliminate the accumulation of long-lived greenhouse gases. The ozone hole was stabilized and continues to shrink.

There are several important aspects to note concerning this phenomenon. It was a clear indication of homogenic activity on a global impact level. We industrial humans partially caused it but we also are underway in solving or reversing the problem. Another crucial element in the problem process relates to the building of consensus within science as a kind of intellectual motor to drive the corrective process.

The ozone hole is perhaps the most dramatic of late-twentieth-century middle-sized problems that have been addressed with some success. Smaller, more regional successes in the reversal of environmental degradation may also be pointed to. The Thames, which underwent cleanup processes helped by legislation in mid century, now contains fish as far upstream as London, which had not been reported in more than 100 years. Similarly, today there are signs posted along streams near my summer home in Weston, Vermont, showing fisherfolk how to identify the differences between trout and young salmon. Salmon, just beginning to recolonize these streams, are still under protection, whereas trout are not so scarce or endangered.

Improvements in the amount of atmospheric lead in North America have occurred, as has the technology of more cleanly run combustion engines. One indicator is that the most recent move to transform two-stroke engines (used in outboard motors, lawnmowers, and chain saws) shows how large engines (such as 200-horsepower auto engines) are many times cleaner than two-horsepower chain saws. Even woodstoves, which used to be at most 35 percent efficient and put out much particulate matter, today reach 75 percent efficiency with little particulate matter.

Nor should we despair that such moves have little or no effect. There is an interesting evolutionary indicator concerning air quality. As early as the eighteenth century in England, and then in the nineteenth century in Michigan, the peppered moth began to turn from a very light-colored moth with few pepper marks to a much darker and highly peppered one in response to trees darkened by soot and industrial waste produced by factory processes. Today, these moths have returned to the light versions as trees turn lighter due to cleanups of the air.

All of this is at least an indication that reversibility can be attained in environmental areas. But I should like to point to a subtheme that is also important the solutions I have been pointing to do not entail either abandoning technologies or pulling plugs. Instead, they often point to improved and higher technologies.

Air conditioning systems using CFC substitutes have been redesigned with higher efficiencies than the older machines that relied upon the greater evaporative and cooling properties of CFCs. Similarly, even presumably lowtech woodstoves have reverse draft systems that more efficiently combust oxidation gases. My new Subaru Outback is peppier than my old Buick Roadmaster of college days, gets more than three times the mileage, uses lead-free gasoline, and has all-wheel drive. In short, the solutions to techno-environmental problems that have worked call for better technologies rather than older, simpler, or no technologies.

Though I am far short of advising that high-tech solutions automatically solve the problems, I am suggesting that retroactive romantic returns to previous low-tech or simpler solutions are unworkable forms of environmentalism. Take solutions from whence they come.

One of the most controversial enduring biotechnological problems in the United States is the deeply emotional one concerning abortion. Abortions are, of course, only indirectly environmental, in that abortion can be one method of population control (as it is in Japan and Russia). Without examining the agonizing psychological and ethical problems associated with this process, one can note that the dominant form of therapeutic abortion in recent times has been the suction method used in the first trimester, usually performed as a sort of industrial version of biotechnology: the abortion clinic. Here, not unlike factory systems, the experts are gathered, time-motion studies for fast performance are entailed, and something like an assembly line occurs.

Again, following the industrial metaphor, highly visible factories have been easy targets for strikes, just as abortion clinics are the targets of abortion opponents. A change of biotechnologies (in this case, the introduction of biochemical day-after or day-after-the-day-after pills), when more fully implemented, will clearly change the way abortion can occur. A much more private client/doctor relationship, perhaps even followed by over-the-counter processes, will soon render the factory strike metaphor, which advantaged protesters, obsolete and create a more decentralized, private occurrence, which will be much harder to make public and much more like miscarriages or spontaneous abortions, which are not condemned by abortion opponents.

None of the examples I have used are global as such. But I have shown how techno-environmental changes are something of a "moving lunch" in which there can often be quick regional or problem changes. Similarly, the variants of "think globally, act locally" today are often quite ambiguous with respect to ecological issues. Reverting again to a New England example, today most New England states have much more forest than at any time in the past 200 years. Along with recovered forest land, long-absent species have returned (in the last two summers I have watched a wild turkey flock develop, and this year a young bull moose took a swim in my pond, both examples of wildlife not even present 20 years ago), and local practices that disfavor clear-cutting remain in place. Maine, the last corporate hold-out, has had to yield to a trimming back of property-holder rights as a result of a 1998 referendum.

These recoveries, however, are often at cost to local preferences in that dairy farming continues to fail and local residents must convert to service industries (caretaking; lawn and meadow care; selective lumber arrangements; maple syrup mining, and tourism), changing or degrading previous forms of rural or agricultural lifestyles. Nor have these changes, which today apparently favor both wilder forests and wildlife, totally eliminated climatological enemies such as acid rain (from Rust Belt areas—a far greater enemy to the environment than we landholding flatlanders), the warming change (which makes it harder for some species of trees to maintain health), or the invasion of televisual culture through minisatellites that threaten older lifestyles more than many other invasions. The ironic result is that old-timer hunters from Vermont, despite statistics reporting abundant deer and moose, take their wilderness hunting to Canada and leave our environs to the flatlanders from Connecticut, New York, and New Jersey. Again, the techno-environmental mix is one of high ambiguity and mixed results.

Moreover, these ambiguities also call into question the usual folk wisdom of many in both philosophy of technology and environmental philosophy, at least with respect to the advice that arises from the various forms of nostalgic, retroactive, or smaller-is-better romanticisms. I have been suggesting that there can be higher-tech solutions to lower-tech problem contributions. I have been suggesting that not all local, regional, or traditional solutions either solve the problems or are necessarily the best solutions. Above all, I am suggesting that technoenvironmental problems in the late twentieth century are almost all complex, interrelated, ambiguous problems that rarely yield to quick fixes or easily grasped solutions.

Critical Realism and Techno-environmental Problems

It is now time to take a concluding look at the hard problems.

I have not meant, in any way, to distract us from facing what I have elsewhere called a foundational problem for the philosophy of technology, that is, the solution to environmental problems (Ihde 1990, 1993). What I have intended is to point out that solutions to these problems must take shape at the appropriate levels of complexity and in the contexts of our technologies. In this conclusion, I shall look not at how we can "win the war," because I do not have any quick fix for that long-run strategy, but instead at some tactical areas where small battles have been won and at the placement of personnel and developments that may suggest tactics that work pragmatically.

I return to recycling, a presumed local solution that addresses part of a problem. Ironically, recycling is not local. Indeed, massive recycling is performed in China, which is the recipient of a lot of American and Japanese junk. The Chinese have complained that the Americans, in particular, are sloppy with their separations and that too much garbage gets into the material. Recycling, like much else in the late modern world, is a multinational process. There is little, if anything, that is limited to the local.

Our industries ship goods to lower-cost labor areas to construct, so we send our recycling to lower-cost hand-labor areas for separation and processing. But the high-tech Germans may come to the rescue with a high-tech solution to the Chinese complaint. They have developed a version of the old cream separator for recycling that will eliminate the hand labor. (Turn the crank and the high-speed centrifuge separated the milk from the cream.) This device simply takes all recycling products and melts and crushes them down into a semi-liquid. Through heat and other processes, the resulting paper residue comes out one spout, glass another, plastics yet another. (Luddites, prepare to protest de-skilling!)

This process, indirectly, points up the hardest nut to crack: How does one turn multinational corporations green? Again, I have no quick fix, but in a few cases there are actually some positive indicators. At a conference, I heard a paper on how several high-tech companies have taken green directions. Two printer manufacturers were stimulated by studies within the companies on recycling processes of toner cartridges that actually saved them money. These promising initial processes led to larger changes. For example, the old process of shipping machines entailed large cardboard containers, volumes of Styrofoam peanuts, and wooden pallets, few of which were recycled. A change to transparent thermal heat wrap, with a deposit return on the pallet, has not only saved hundreds of dollars but had an interesting positive side effect. Handlers, actually seeing the fragile machines, seemed to take more care than in the older cardboard box days, and shipping damages went down. The point of this example is that when green processes can be demonstrated to produce lower costs or contribute to higher profits, corporations will adapt accordingly. It takes innovators and pushers to develop this green efficiency. I am convinced that green high-tech processes can do precisely this.

What role does or can the philosopher play in this green turn? For a long time, I have argued that our usual role is one doomed to reactive status.

This is particularly true in the case of applied ethics, which began with medical contexts and spread to business schools, law, and other contexts, where the presumption is that technologies dispose, and it is up to us to make that disposition as ethical as possible. I do not want to unemploy all those philosophers who have made livings doing just that, but I prefer a more proactive position.

The proactive position that I am advocating places the philosopher at ground level, particularly in the research and development phase of technological processes. For a while I despaired of this happening, but in recent years I have observed and even participated in several of these projects. Paradoxically, some of these occurrences have happened precisely because of the displacement from traditional academic vocational roles. In the United States, for example, some philosophers, lacking full-time academic placement, have found themselves freelancing part-time in computer companies or other high-tech companies. Here the kind of disciplinary thinking for which we have been trained comes into play in a new situation. Critical thinking can be applied to developmental, not resultant, processes. Contributions to some of the green directions just mentioned have been helped along by precisely such part-time philosophers.

More interestingly, in the last few years I have been doing quite a bit of commuting to northern European technical universities, where I have found increasing numbers of retooled philosophers, particularly philosophers of science, working not in traditional departments of philosophy, but within these polytechnics. They often become part of interdisciplinary research teams of engineers or other applied sciences. These kinds of situations are challenging and cast a very different perspective upon philosophical application. I am suggesting that, at the very least, this front-loaded position should be one for engagement just as much as the end-loaded situation our ethicists have found themselves within.

I do not want to conclude with any easy sense that all techno-environmental problems are simply fixable by applying more technology or that the removal of technologies does not solve the problems we now have. I want to indicate that all techno-enviromental problems are complex, ambiguous, and interwoven. The tasks are not easy, but neither utopian nor dystopian attitudes ultimately help. It is precisely within the ambiguities and complexities of our current situation that philosophers must take their places, and it should be obvious that techno-environmental problems take their very shape at these locations. What better place to be than the twenty-first century?

References

Abram, David. 1996. *The Spell of the Sensuous*. Pantheon Books.

Brown, Lester R. 1996. *State of the World: A World Watch Institute Report on Progress Toward a Sustainable Society*. Norton.

Ihde, Don. 1993. *Philosophy of Technology*. Paragon.

Ihde, Don. 1990. *Technology and the Lifeworld*. Indiana University Press.

3 Is Technology Use Insidious?

Kyle Powys Whyte, Ryan Gunderson, and Brett Clark

The philosophy of technology and environmental philosophy converge on the topic of the *insidiousness* of technology. Insidiousness is the idea that communities that adopt technological hardware, such as televisions, or rely on technological supply chains, such as global agrifood, cannot stop the erosion of their previous, more intimate relations with the environment. In both fields, many see the erosion as bad for various reasons, as it undermines taken-for-granted relationships with non-human species and ecosystems, as well as family and other social and cultural relationships that sustained ways of life that were less dependent on complex technologies. It creates the illusion that humans are not dependent on and cannot be affected by Earth systems such as the climate system. It alienates people from the combined impacts on the environment of their individual actions, such as the accumulation of greenhouse gases in the atmosphere. Aldo Leopold (1949, 12), whose ideas intersect with both the philosophy of technology and environmental philosophy, addressed these concerns when he wrote: "There are two spiritual dangers in not owning a farm. One is the danger of supposing that breakfast comes from the grocery, and the other that heat comes from the furnace."

In this chapter we compare and contrast notions of the insidiousness of technology in the philosophy of technology and areas of environmental philosophy concerning environmental ethics and normative environmental theory. We begin by discussing the opposite of insidiousness, technological neutrality, and disinterestedness, as these concepts influence the arguments that follow. We then examine perspectives from the philosophy of technology and environmental ethics that argue that technology is not neutral or disinterested because it embodies the particular lifeways of the societies that produce the technological tools in question. We consider the implications of this argument as it relates to changes in lifeways, values, and intimate relationships. We then present the division within normative environmental theory regarding questions related to technology. We discuss the technological optimism evident in the ecological modernization perspective as well as the critique of technology offered by environmental political economists (who argue that technology is not neutral and that it is insidious, given the social relations of the capitalist system).

We close with a discussion of the argument that technological rationality furthers the logic of capital, extending the insidiousness of technology, under the modern industrial system. This discussion highlights a major division in regard to technology and illuminates the range of questions, issues, and relations related to assessing technology and its consequences. We suggest that a monolithic understanding of technology can be problematic, as a more nuanced consideration of the social and historical relationships can be informative. The more nuanced approach acknowledges the full force of insidiousness without erasing the possibility of certain limited forms of resistance to capitalism using technologies. We conclude with a set of questions that seek to bring out insights regarding how technology, and by extension the various social relationships that shape it, transform our intimate relations with the environment.

The Neutrality View of Technology

A major debate within the philosophy of technology concerns whether technologies exert influence on individuals beyond the technologies' intended uses. We do not aim to address all the twists and turns of that debate. Instead, we limit our brief discussion to concerns associated with neutrality and intention, as they are central to the issue of insidiousness.

What happens to people when they adopt technologies? Neutrality is the idea that technologies are just physical things that affect people solely according to what they intend to do through application of the tools. According to this assumption, technologies are simply things lying around, waiting to be put to use. This idea communicates two claims. First, humans can design technologies to do whatever they want. Second, humans can use technologies in any way they please regardless of the original intention behind the design of a technology. These claims suggest that the outcomes of technological design and use are completely understandable in terms of what humans do. Furthermore, technology contributes nothing in its own right to any aspect of its entanglement with humans.

The two claims, then, establish that the best way to explain the outcomes of human technology use is in terms of human intentions and practices. This approach does not preclude technologies from having unpredictable outcomes or simply not working according to plan. In fact, unpredictability and malfunction result from human ignorance, errors, and limitations. There is nothing about a technology that makes it more or less prone to such outcomes beyond considerations regarding how humans tried to make a device or ended up using it. According to Andrew Feenberg (2003), this view makes technology appear

> purely instrumental, as value free. [Technology] does not respond to inherent purposes, but is merely a means serving subjective goals we choose as we wish. For modern common sense, means and ends are independent of each other. Here is a crude example. In America we say "Guns don't kill people, people kill people." Guns are a means which is independent of the ends brought to them by the user, whether it be to rob a bank or to enforce the law. Technology, we say, is neutral, meaning that it has no preference as between the various possible uses to which it can be put. This is the instrumentalist philosophy of technology that is a kind of spontaneous product of our civilization, assumed unreflectively by most people.

The neutrality view, then, suggests that technologies, on their own accord, exert no preferences, values, or prescriptions that contribute to whatever affects the technologies have on users.

The view of technology as neutral can be coupled with another view, which we call the disinterestedness of technology. This view suggests that when people use technologies, they are not affected in any additional ways beyond the enhancement or decline of their abilities to do what they desire to do. That is, technology either enhances according to plan or fails to enhance according to plan. There are no profound changes occurring in human users attributable to the use or presence of various technologies. To be clear, while users' aspirations may be heightened when a technology makes feasible something that previously was not feasible (e.g., commercial space shuttles), the technology operates on people's preferences in perfectly straightforward ways. Users would have aspired to space travel had the technology for doing so been available; the technology for doing so now exists. This idea is also present in debates about how television affects people. For example, television itself does not make people have less self-esteem, become obese, and lose manual dexterity. Rather, television merely fulfills some people's propensities for these outcomes. People are not forced to watch television and can eliminate it from their lives if they want. Moreover, television could be changed to be less likely to fulfill these propensities. The point is, following the logic presented above, that it is the human propensities that matter, not television.

Christian Illies and Anthonie Meijers (2009) propose a particular form of disinterestedness in their concept of action schemes, which refer to the array of possible actions that are available for someone to perform, including the degrees of attractiveness of each action relative to the particular individual. Everyone has action schemes that explain part of what it means to make choices and do something. The design of technology affects how these tools change peoples' action schemes. For example, having access to cell phone apps may make it more attractive to order delivery from a restaurant instead of cooking or to take a taxi home instead of walking or taking a bus. If these expenses put someone in debt, we do not blame the phone apps, but the person. So the apps do not participate in any profound way in the person's agency beyond simply making certain kinds of choices more or less attractive. In this case, though the technology does change something close to the person adopting it, the effects can be understood straightforwardly. Illies and Meijers use the classic example that when someone murders another person with a pistol, the pistol is not accused of murder. In other words, it does not change fundamentally who the person is or his or her personality or what the person can be blamed for. It simply changes how actions are presented. This change in possible actions has ethical implications: Some technological interventions into action schemes may harm users; others may enable them to do better things with their lives. Thus, engineers and scientists ought to consider how their designs and inventions affect people's action schemes.

Philosophy of Technology and the Insidiousness of Technology

Numerous philosophers argue that technology is neither neutral nor disinterested. Human design and practice are not the only variables that explain the outcomes of technology use and how users may change as a result of adopting a technology. In regard to the "guns don't kill people" example, Don Ihde (1990, 27) argues that "it becomes immediately obvious that the relations of human-gun (a human with a gun) to another object or another human is very different from the human without a gun. The human-gun relation transforms the situation from any similar situation of a human without a gun. At the levels of mega-technologies, it can be seen that the transformational effects will be similarly magnified." His point is that the significance of an individual having a gun in a situation is not reducible to one's carrying a weapon for security purposes just in case something extremely unfortunate happens. Having a gun actually changes the dynamics of the situation in ways with which humans cannot fully come to terms in relation to their own propensities. An individual takes on a different identity, power relations shift, and social atmospheres change. There is, then, something distinctive about the gun, and what it does to the situation, that was certainly not intended by its designers and which is not wholly under the control of humans embedded in a particular situation. Bruno Latour (2009, 159) describes this *something* as a "third possibility," or an

> uncertainty, drift, invention, mediation, the creation of a link that did not exist before and that to some degree modifies the original two [the gun and gun holder]. Which of them, then, the gun or the citizen, is the actor in this situation? Someone else (a citizen-gun, a gun-citizen). You are a different person with the gun in your hand. ... This translation is wholly symmetrical. You are different with a gun in your hand; the gun is different with you holding it. You are another subject because you hold the gun; the gun is another object because it has entered into a relationship with you.

The presence of the gun, then, transforms the situation profoundly. That is, the human user becomes someone different in relation to others; the gun becomes a different kind of technology than it may have been before it was introduced into that situation. The result is not just a new person or a new gun, but the emergence of a mixed identity that cannot be reduced to the aggregation of the original two identities. The two interact with and mutually constitute one other (Latour 1993). As a result, the transformation is more profound than just a gun fulfilling or failing to fulfill the initial human propensities that led to its adoption.

The non-neutrality views of Ihde (1990) and Latour (2009) complement the arguments of feminist philosophers of technology. Science and technology often embodies social relations, which include oppression, power, and inequality (Haraway 1991; Harding 1991, 1998). These "powerful background beliefs" influence the bounds of human understanding of social relations, including those with technology, whether intended or unintended (Harding 1991, 149). For example, Val Dusek (2006) discusses the complex social relationships and the shift in knowledge that accompany the use of fetal ultrasound imaging. This technology provides a better image for the doctor in order to render better medical service, but it also changes the power dynamics of the situation. Fetal ultrasound imaging privileges visual knowledge over bodily knowledge, giving the physician greater insight and control over the process. The ultrasound image also presents the fetus in a particular way that endows it with an identity it could not have when it was not visualized. The fetus, in the image, appears independent, removed from the context of the mother's body. The technology changes the dynamics of this social relationship and changes the general perceptions of a fetus. It also changes what assumptions people make about the situation and how they perceive their identities and those of others. In the case of ultrasound imaging, the fetus takes on a new identity that it previously did not have. There is nothing in the intentions and designs of ultrasound equipment that would have suggested that this would occur. Feminist philosophers of technology suggest that understanding these dynamics—oftentimes through hindsight—requires an assessment of the larger social context and the power of images within society at a particular historical point in time.

The argument that technologies are insidious is a version of the nonneutrality view of technology. Albert Borgmann (1984, 1999) exemplifies this tradition. In his historical analysis of technology, he discusses how, in the pre-industrial world, technologies were greatly limited in the particular benefits that were provided, such as heat. The limitations meant that people had to exert much skillful effort to extract these benefits from the technologies. In regard to the traditional hearth, Borgmann (1984) explains that people had to be experts in both the extraction of the resources needed to fuel the hearth and in the management of fire for cooking and heating. All of the tasks surrounding the use of a hearth could not be performed by one person, but required a family. The limits of the hearth's ability to heat a house resulted in an additional benefit, as social life in the household was centered around it. The hearth required multiple skills and served multiple functions, from heating to cooking to socializing. A hearth suggests a community whose members have particular intimacies with their fellows, the environment, their sustenance activities, and the materials they use to maintain their ways of life. The limitations of technologies are important sources of these intimacies, which served as a significant part of their survival and connection with the larger biophysical world.

A similar situation is evident in Amish communities. In contrast to a widespread notion, the Amish are not anti-technology. They employ an array of technologies within their fields, shops, and homes. The adoption of specific forms of technology involves extensive social deliberation, contemplating what the potential consequences will be for the Amish way of life and their communities. For example, television is often seen as a technology that would decrease family cohesion and conversation. West-facing windows maximize radiant heat and encourage family members to spend time together in specific parts of the house. The design and the use of technology involve social relations and the types of solidarity that may arise.

Borgmann (1984) argues that particular changes in technologies in industrial societies have had real and lasting consequences, which are insidious. The transition from traditional hearths to central heating systems, microwave ovens, and televisions ushered in a series of changes. People no longer have to possess skills and knowledge, especially in regard to their environment, which helped sustain previous generations. Relationships with the larger community are transformed, as direct reliance on each other is no longer part of everyday existence. Central heating systems provide as much warmth as people want with a flick of a switch. There is no skill involved. People do not have to know what resources, such as coal energy, go into the warmth provided by a central heating system in order to use it. They certainly do not need to rely on others on a daily basis. If the central heating system breaks down, they can call a specialist to come and fix it. Central heating provides only one benefit: warmth. It cannot be used for cooking, and it is not a basis for socializing. As a result, people can get by without knowing how the heating or microwave technologies work and what environmental inputs are required to supply power. These technologies can be used individually without having to be entertained by anyone else. Borgmann (1999) contends that technologies, such as central heating and television, have an insidious effect on communities that previously did not rely on these technical devices:

> Only about a quarter of the people in this country and 1 percent of the world's population are affluent enough to own a personal computer, have access to a computer network, and need to worry about e-mail flirtations and CD confusions. Yet the less affluent and less educated citizens of the United States are drenched with information as well. Television is the major channel saturating them with news and entertainment. Though they are more passively connected to information, their connection to reality too is profoundly transformed. The breathless glamour of television numbs their ability to confront and endure the gravity and pressure of reality. Information is the element of technological affluence that invades the culture of poor and premodern countries most quickly and easily. First come the transistor radios and then the television sets, the latter few in number but watched by many. If information is not the medium of an overwhelmingly new culture, it is at least the entering wedge that permits indigenous cultures to seep away and disappear. (Borgmann 1999, 6)

Borgmann's point is that the adoption of technology is insidious—that it erodes important intimate relations within communities and with the environment. That is, technology adoption transfers the values, ideas, and lifestyles of the society that designed the technology, and the adopting community is helpless to respond.

The view of technology as insidious contrasts sharply with the notion of neutrality and even more so with the notion of disinterestedness. According to this position, there is something profound that occurs when humans adopt the technologies of other societies. This effect goes far beyond simply changing action schemes and potentially has significant social and environmental consequences.

Environmental Values and Insidiousness

The insidiousness of technology appears in environmental philosophy in work on environmental ethics, most often in relation to Indigenous communities, but also in regard to Amish and other traditional religious communities. J. Baird Callicott has written a series of articles and a book comparing his interpretation of Aldo Leopold's land ethic with his research on the lifeways of specific Ojibwe communities in the Great Lakes region. He has developed an approach for exploring the historical ethics of a particular group of people and their culture by examining their language in order to determine how they related to physical objects, non-human living things, other people, and environmental systems. He suggests that the ethics of certain Ojibwe communities provide much insight for a useful comparison with Leopold's land ethic. For example, Callicott discusses how Ojibwa ethics are based on understanding nonhuman plants, animals, and entities as fellow family and community members. Leopold too referred to the origin of ethical responsibility as based on seeing oneself as a member of a biotic community. In presenting this work, Callicott (1989) describes how the colonization process has influenced the lifeways of Ojibwe and other Indigenous communities. In particular, he argues that the integration of technologies developed by colonial societies has had an exceptional impact on tribal communities. He claims that "to adopt a technology is, insidiously, to adopt the worldview in which the technology is embedded" (ibid., 36). He explains:

> Modern Indian life (except perhaps among a few Amazonian peoples) reflects the inescapable influence of global technological civilization and of the ideology that engendered it and in which it remains grounded. This influence is not only inescapable; it is also insidious. Technologies are never cognitively and axiologically neutral; they are embedded in an engendering and sustaining system of ideas. To buy guns, motors, and mackinaw jackets is to buy, however unintentionally, a worldview to boot. Both the Koyukon and the Mistassini Cree use guns and steel

> traps in taking game, wear some store bought clothes, and, to one degree or another, Christianity, modern science, Western medicine, money, and materialism affect them. (ibid., 39–40)

For Callicott, technology is inevitably associated with the dominant worldviews of the society that created that particular technology. In other words, worldviews, cultures, and ideologies are embedded within technologies. All of these things can be seen as lifeways. These lifeways are not only embedded within the technologies themselves, they are also *transferable* once individuals use the technology, no matter what their purpose for using the technology is. That is, for Indigenous persons, an individual may adopt a technology to simply hunt more efficiently, which is the person's explicit purpose. But in adopting the technology within the context of a previous lifeway, then that Indigenous person will inevitably change the ideas that he or she has about previous lifeways and become a new person. The more community members engage in this activity, the more the community as a whole inevitably changes.

This idea is worth exploring in somewhat more depth. The assumption follows this logic. Before an individual had access to store-bought goods, he or she may have had the idea, embedded within the lifeways of the community to which the individual belonged, that consumption of such items was not meaningful. As a result, the person did not participate in the sorts of consumptive practices characteristic of a culture that makes store-bought items. The store-bought items themselves, in some way, contain the ideas and values of the society that produced them within their very substance. When these items are physically used by people in a different culture, the substance of the goods—i.e., the lifeways of another society—becomes an integral part of the interaction. The sustained physical contact between the store-bought objects and the person results in the transfer of outside lifeways into the new community, eroding important intimate relations that previously helped define and sustain the individual. Hence, the Indigenous person now becomes much more materialist and seeks more consumption once he or she begins to acquire store-bought items. The effects are certainly greater the more that person adopts the technology.

Citing an example very similar to Callicott's example of Indigenous peoples, Borgmann (1999, 25–27) writes:

> The ancestral environment of the Salish was well-ordered as well as coherent because some natural signs stood out as landmarks from among the inconspicuous and transitory signs of creeks, rocks, trees, and tracks. Landmarks were focal

> points of an encompassing order. ... Just as signs and things were naturally and intimately related in the ancestral environment, so were information and knowledge. To recognize a sign was to know what it meant. ... So to be intimate with one's world must have been a pleasure when people were healthy, food was plentiful, and conditions were peaceful. ... Pleasure and pain have not been dissipated by technological information, but they have become focused and distracted through the ubiquitous intrusion of signs into the presence of things. By now we are so inured to the blight of untrammeled information that it takes a deliberate withdrawal to something like the ancestral environment if one is to notice the damage done. Bill McKibben did so by counterposing twenty-four hours of television to twenty-four hours in the natural setting of the Adirondacks. ... Television advertising constantly abets our belief that ever new bits of property can make up for our failure to appropriate the focal area of our lives. Here the ancestral world offers a salutary contrast. ... Being out in the wilderness restores one's sanity and serenity.

Parallel to Callicott's argument, Borgmann views information technology as insidiousness. The adoption of foreign technologies, especially on Indigenous groups, erodes their culture without there being much they can do about it. Ecophenomenologists, using different traditions of phenomenology and environmental ethics and usually not focused on Indigenous peoples, have made comparable cases about technology (Abram 2005; Brown and Toadvine 2012; Wood 2001).

These views are supported by some environmental research on values and ethics that addresses traditional ecological knowledge (TEK). Though the following definition is debated (Whyte 2013), TEK often refers to the accumulation of intergenerational knowledge of natural systems and cultural interactions with the biophysical world. This knowledge is passed on from one generation to the next, enabling Indigenous communities to sustain themselves in situations where they depend heavily on how they are able to respond to the environment (Berkes 1999). TEK is particularly important for Indigenous populations because it is the basis of how they make plans for the future and respond to environmental changes, such as shifts in the location of animal populations and in weather patterns. TEK also has important value and belief aspects. Knowledge is associated with ethical practice toward the environment and with religious beliefs. A particularly important aspect of TEK is perception. Communities that use TEK do so in expert ways. That is, TEK is not an algorithm for interpreting environmental change. Rather, it is an ability to perceive environmental changes that would not be obvious to someone without experience in that particular environment. Particular types of scientific expertise, then, can be said to fall short of the perceptual abilities of TEK.

Lilian Alessa and her colleagues conduct research on how environmental perception changes with technology adoption. Environmental perception is important because there is a difference between what community members perceive to be the amount of a given resource and the actual amount of the resource. The same is true as far as the perceived changes in an environment and the reality of the changes. The closer perceptions are to reality, the better communities can adapt to changes in climate and to other environmental changes. However, the adoption of technology can alter this relationship. For example, Alessa et al. (2010) suggest that a phenomenon known as Technology-Induced Environmental Distancing (TIED) is occurring in Indigenous communities that adopt certain types of technologies. As part of their analysis, they explain:

> [We] provide empirical evidence to suggest that water users in rapidly modernizing, remote, resource-dependent communities are experiencing TIED. Associated with TIED is a shift in worldview, as expressed values, from that of an indigenous cultural identity (ICI), where cultural and subsistence values are most important, to the dominant social paradigm (DSP), where individual benefit and convenience emerge. Values help explain how individuals behave, and they affect and influence the broader macro-level patterns, rates, and norms by which people live. … We posit that the values of water shift from more holistic to more convenience-oriented values in users who have become distanced from the totality of freshwater as an integral part of the social-ecological system in which they live. Distancing is defined as any physical or cognitive barrier between people and a resource that leads to a diminished awareness of that resource. … A decreased awareness of the resource, that is, desensitization, is therefore a consequence of the increased distance from the resource. This encumbers trade-offs between short- and long-term resilience. (ibid., 255)

Alessa et al. claim that the adoption of certain technology facilitates the shift to the dominant worldview, which includes lessening previously important environmental values.

Research on environmental ethics and values suggests that the adoption of technology affects people's attitudes toward the environment or the degree to which they value it. The implication is that this influences people's ethical conduct, insofar as values are often understood as playing a role in what individuals think is right or wrong. As a result, the adoption of technology plays a role in eroding social relationships with the environment. Technology, then, is insidious.

Environmental Sociology, Neutrality, and Insidiousness

Within the social sciences, including environmental sociology, there are widely divergent views in normative environmental theory regarding technology. These views are comparable to the views discussed in the philosophy of technology and environmental ethics. Here we simply discuss two of the distinct perspectives within this field that address the relationship between technology, society, and the environment. Each view offers a different assessment of modernity and the consequences of technological innovation within the dominant social system.

Normative environmental theories, along with popular beliefs, generally see technological development as a progressive force. In other words, technology is an important part of the "modernization" project. It serves as the means to provide endless bounty and to free people from useless toil. Technological optimists generally assume that technological innovations and breakthroughs will allow society to overcome socio-ecological challenges and natural limits (e.g., shortages of fossil fuels will be addressed through development of green energy and improvements in energy efficiency). Julian Simon (1981) famously argued that there were no serious environmental problems or natural constraints. According to him, the capitalist market encouraged technological innovations, which would support endless economic and population growth. More recently, Bjorn Lomborg (2001, 2008) has suggested that technological development allows humans to improve the world, rather than harm it. Lomborg argues that global climate change is not a serious concern, and that further economic and technological development will serve as the means to address any concerns that arise. Technological optimists generally take for granted that technology can do whatever we want and can be applied as we please. They presuppose that technology is both neutral and disinterested. They view environmental problems as technical problems that require technical solutions or technological fixes.

Within environmental sociology, ecological modernization theorists tend to adhere to a technological-optimist position. They propose that environmental problems are unavoidable during the early stages of modernization, as economic growth comes at the expense of ecological degradation. Nevertheless, modern capitalist society has reflexive capacities that allow the social system to self-correct and to pursue environmental sustainability. The forces of change include culture, market, the public, and technology. It is suggested that "environmental concerns" become a central part of private-sector decision-making, which generates technological innovations to create green productive practices and products (Mol 1995, 2002; Mol and Jänicke 2010; Mol, Spaargaren, and Sonnenfeld 2010).

Ecological modernizationalists claim that technological innovation throughout the productive system will allow industries to respond to environmental problems, rather than to rely on "end-of-pipe technology" (e.g., smokestack filters), which reduces pollution only after it has been produced (Huber 2010, 18). Arthur P. J. Mol (1997, 141) argues that "environmental improvement can go together with economic development via a process of delinking economic growth from natural resource inputs and outputs of emissions and wastes." He argues that technological innovations and social-environmental reforms could culminate in "an absolute decline in the use of natural resources and discharge of emissions, regardless of economic growth in financial or material terms (product output)." Likewise, Joseph Huber (2010, 335) contends that technological innovations reduce "quantities of resources and sinks used," whether measured by intensity, per unit, or "absolute volumes."

Technology serves as the "linchpin" for the ecological modernization argument (Fisher and Freudenburg 2001, 702). As Mol (2001, 58) argues, "environmental deterioration is conceived of as a challenge for socio-technical and economic reform, rather than the inevitable consequence of the current institutional structure." For proponents of ecological modernization, technology is not insidious. In fact, technological diffusion and adoption are seen as the primary means to create a more sustainable society.

Political economists, including some within environmental sociology, offer a more critical assessment of the role of technology in capitalist development and how it contributes to human-nature relationships. Environmental problems are not seen as technical problems that can be addressed simply via technical solutions. Instead, these problems are seen as arising from the organization of human society, and from its interaction with the larger biophysical world (Foster, Clark, and York 2010; York and Clark 2010). For political economists, the social and historical context of technology is of utmost importance. They propose that the technological-optimist belief that technology itself can solve environmental problems, absent changes in the social relations, is shortsighted.

John Bellamy Foster (2000, 200–201), an environmental political economist, points out "that the Greek word 'organ' (*organon*) also meant tool." Similarly, Karl Marx conceived of technology as an extension of human beings and as something used to interact with nature. Technology is not neutral, as it is developed through social production. Marx explained that technology, while not fully determined and constrained by purely capitalist desires, is often developed and employed to service the interests of those in power—such as facilitating the division of labor or increasing production through the exploitation of labor and nature. Technology embodies capitalist relations, but does not absolutely determine the outcomes. Marx noted that economic conditions determine the abuse and "misuse" of "certain portions of the globe" (1991, 753). "The contemporary use of machines," he wrote, "is one of the relations of our present economic system, but the way in which machinery is utilized is totally distinct from the machinery itself. Powder is powder whether used to wound a man or to dress his wounds" (Marx and Engels 1975, 33). Thus, environmental political economists propose that the influence of technology varies. Technology is not monolithic or one-dimensional. At the same time, technology embodies power and facilitates change. The historic development of technology represents, in part, changes in the relationship between humans and the larger biophysical environment.

Environmental political economists, such as those associated with the treadmill-of-production approach, argue that the growth imperative of the capitalist system results in an "enduring conflict" between capitalism and the environment (Foster 2000; Schnaiberg 1980). They argue that technological innovations play a crucial role in capitalist development, rationalizing labor processes, generating cost reductions via automated production, and enhancing the exploitation of resources (i.e., matter and energy). Allan Schnaiberg (1980), a treadmill-of-production scholar, distinguishes between lower-risk and higher-risk technologies. He explains that lower-risk technologies were developed in pre-industrial and early-industrial societies. These technologies tend to cause environmental problems through "historical accumulation" or through "high-volume" use. For example, the environmental problems caused by extractive industries, such as coal mining, often cause extensive ecological degradation of land and over-exploitation of resources. He points out that low-risk technologies continue to be present in advanced capitalist societies. Higher-risk technologies arose with advanced industrial societies. These technologies present great risks even at lower rates of application. For example, Rachel Carson (1962) and Barry Commoner (1971) showed how the Chemical Revolution introduced synthetic chemicals that created new environmental problems. The use of pesticides in agriculture contributed to the bioaccumulation of toxic chemicals throughout the food web. Given the growth dynamics that accompany capitalist development, high-risk technologies have become more widespread.

Schnaiberg (1980) explains that the development of high-risk industrial technology created additional environmental concerns. First, the development of late-industrial technologies often increases energy inputs to produce and distribute commodities. More energy-intensive machinery is employed to displace human labor. Additionally, energy-intensive materials, such as plastics and chemicals, are incorporated into the manufacturing process. These materials also generate more waste and pollution (Foster 1994; Gould, Pellow, and Schnaiberg 2008; Pellow 2007; Schnaiberg and Gould 1994). And the materials used in production and the waste produced is qualitatively different and presents new dangers. Synthetic chemical wastes do not become less toxic through decomposition, and they can be transferred through ecosystem species and present new health risks (Carson 1962; Steingraber 2010).

This argument stands in sharp contrast to the technological optimism of proponents of ecological modernization. Thus, it is useful to consider the relationship between technology and energy efficiency. Environmental political economists recognize that new technologies often make using energy and raw materials more efficient, but they contend that innovation does not necessarily dematerialize society or contribute to an absolute decoupling of economic development from energy and resources. Referring to the "Jevons paradox," they explain that more efficient resource usage often increases overall consumption of that particular resource. As a result, expanded production outstrips gains in energy efficiency (Clark and Foster 2001; Jevons 1906; Polimeni et al. 2008). They suggest that efficient operations produce savings that expand investment in production within the larger economic system, and thereby increase consumption and total energy consumed, raw materials used, and carbon dioxide produced. They explain that the use of technology must be situated within global capitalism's overall social relations and dynamics. The growth imperative, as suggested by treadmill-of-production theorists, is geared to maximize throughput of energy and matter, thus conservation does not take place at the macro scale of the economy. As a product of capitalism, the Jevons paradox illustrates that purely technological means cannot solve ecological problems such as climate change (Foster, Clark, and York 2010).

If late-industrial technologies are more risky and have qualitatively different impacts on ecosystems and humans, why are they so pervasive? Schnaiberg (1980) provides two answers to this question. Technology is not neutral and it is insidious, given its development within the capitalist system. In regard to the non-neutrality of technology, Schnaiberg discusses how corporate investments, especially after World War II, contributed to technological changes to enhance production and profits. The use of synthetic chemicals and the adoption of more energy-intensive machinery and materials played a vital role in this expansion. Industry effectively "incorporated" technological research into production. The concentration and centralization of capital into monopolistic, large-scale industries made room for an "accelerated stream of innovation" through investments in technological investigation and application as a "protection of an expanded future market for large-scale enterprises" (ibid., 126). In regard to the insidiousness of technology under capitalist relations, Schnaiberg (ibid., 121) states that "the discontinuity represented by modern technology is the less visible, often less detectable ways in which ecosystem effects are felt." Whereas the environmental consequences of a strip mine using early-industrial technologies are more obvious, the impacts of synthetic chemicals and nuclear power are often more uncertain and invisible. Schnaiberg (ibid., 132) explains that there are a number of additional reasons for this "social invisibility" of late-industrial technologies: (1) decreased employment in direct production activities, (2) the decentralization of many production functions, (3) especially low employment in environmentally extractive activities, and (4) technological progress viewed as routine, anticipated social change. For example, offshore oil rigs are largely mechanized (i.e., less dependent on human labor power) and quite removed from communities. Similar to several of Borgmann's points, Schnaiberg suggests communities often do not "see" the environmental impacts but instead see the potential social gains (more jobs, etc.). In fact, many communities perceive themselves as dependent upon economic growth to secure jobs and taxes to fund government and social programs. The invisibility of the impacts of technology on ecosystems is "inherent in the nature of these technologies, which are capital-intensive, substitutive of large volumes of raw materials, shipped in the most economical forms of transportation over long distances, and passed quickly through ecosystems into biospheric sinks" (ibid., 133). Within this context, environmental degradation is normalized, and simply becomes a way of life (Buell 2003).

Environmental political economists highlight the private interests that drive technological change and the social invisibility of late-industrial technology. They illuminate the social, economic, and political context that influences technological innovation and deployment. They call into question the technological optimism that is so prevalent among the environmental mainstream and that informs policy. They illustrate the importance of power and how it relates to the non-neutrality and insidiousness of technology as a mediator of relations between humans and nature under capitalism.

Conclusion

A major divide exists within the philosophy of technology and within environmental philosophy (including environmental ethics and normative environmental theory) regarding whether technology is insidious. Proponents of ecological modernization and other technological optimists generally consider technology neutral. For them, technology is not insidious. In fact, technology is a progressive force, enriching social and environmental conditions. Rather than eroding intimate relationships with the environment, technology provides the means to pursue sustainability. Other philosophers of technology, including feminist philosophers of science and technology and environmental political economists, contend that technology is not neutral and that it is in fact insidious. The logic of the arguments varies, as an array of relationships, conditions, and questions are considered. Philosophers such as Borgmann and Callicott propose that technology actually embodies the lifeways of societies that created these tools. When specific types of technology are used by people who reside outside of the societies that created it, the lifeways contained within the tools are transferred to this population. These lifeways transform the culture and intimate relationships, generally eroding important values and ethics that had previously contributed to the well-being and environmental sustainability of these people.

Environmental political economists and sociologists argue that any assessment of technology must consider the social and historical relationships that influence technological innovation and deployment. In a capitalist economic system, a particular type of logic and rationality is imposed on humans and nature designed to further the accumulation of capital. This socioeconomic system contributes to the insidiousness of technology, further eroding intimate relationships with the environment, deepening alienation from nature, and accelerating ecological degradation. Yet, at the same time, there are views and studies that show how groups living in opposition to the capitalist economy have harnessed technologies associated with capitalism and settler colonialism for protecting their land-based ethics, values, and lifeways. For example, Nick Reo and Kyle Powys Whyte (2012) discuss the view that Indigenous subsistence hunters in one community use a range of technologies from rifles to trucks as a way of better connecting with the land, respecting animals, and instructing young people about their responsibilities to animals and the environment. This study and others cited in it do not deny insidiousness in the ways just described in this chapter. However, they do suggest that it is possible to conceive of limited forms of resistance to capitalist and settler colonial systems that can involve the use of some technologies. A related but different example is provided by the Indigenous mass mobilization Idle No More, which uses social media and other technologies as part of strategies that emphasize how Indigenous peoples "cannot live without the land and water. We have laws older than this colonial government about how to live with the land" (Idle No More 2014).

These distinct traditions and views in this chapter reveal that a broad range of relationships and issues are associated with discussions of technology. It is clear that a nuanced approach is necessary in order to capture the social and historical relationships and conditions that influence technology and its impacts on societies. At the same time, it is clear that technology itself acts as a powerful tool that shapes social-environmental relationships. As a result, questions associated with technology must be part of the larger discussion regarding environmental sustainability. Philosophers of technology, environmental social scientists, and environmental philosophers should consider the following questions when examining different technology-society-environment interfaces: What social and social-environmental relations give or gave rise to the given technology and how does the latter reproduce or transform the former relations? Who is in control of the design and application of the given technology? How is the given technology governed? Which social groups and environments benefit from the given technology? Which social groups and environments are harmed by the given technology? Which values, interests, and politics are reflected in the design and application of the given technology? Addressing these questions is important for understanding the social dimensions of technology-social-environment relations and the social and environmental implications of technological development and application. The answers may compel the researcher to question the view that technology is neutral and disinterested.

References

Abram, D. 2005. Between the body and the breathing Earth: A reply to Ted Toadvine. *Environmental Ethics* 27 (2): 171–190.

Alessa, L., A. Kliskey, and P. Williams. 2010. Forgetting freshwater: Technology, values, and distancing in remote Arctic communities. *Society & Natural Resources* 23 (3): 254–268.

Berkes, F. 1999. *Sacred Ecology: Traditional Ecological Knowledge and Resource Management.* Taylor and Francis.

Borgmann, A. 1984. *Technology and the Character of Contemporary Life: A Philosophical inquiry.* University of Chicago Press.

Borgmann, A. 1999. *Holding on to Reality: The Nature of Information at the Turn of the Millennium.* University of Chicago Press.

Brown, C. S., and T. Toadvine. 2012. *Eco-Phenomenology: Back to the Earth Itself.* SUNY Press.

Buell, F. 2003. *From Apocalypse to Way of Life: Environmental Crisis in the American Century.* Routledge.

Callicott, B. 1989. American Indian land wisdom. *Journal of Forest History* 33 (1): 35–42.

Carson, Rachel. 1962. *Silent Spring.* Houghton Mifflin.

Clark, B., and J. B. Foster. 2001. William Stanley Jevons and the coal question: An introduction to Jevons's "Of the Economy of Fuel." *Organization & Environment* 14 (1): 93–98.

Commoner, B. 1971. *The Closing Circle.* Knopf.

Dusek, V. 2006. *Philosophy of Technology: An Introduction.* Blackwell.

Feenberg, A. 2003. What Is Philosophy of Technology? Paper presented at the Lecture for the Komaba Undegraduates.

Fisher, D. R., and W. R. Freudenburg. 2001. Ecological modernization and its critics: Assessing the past and looking toward the future. *Society & Natural Resources* 14:701–709.

Foster, J. B. 1994. *The Vulnerable Planet.* Monthly Review Press.

Foster, J. B. 2000. *Marx's Ecology.* Monthly Review Press.

Foster, J. B., B. Clark, and R. York. 2010. *The Ecological Rift.* Monthly Review Press.

Gould, K. A., D. N. Pellow, and A. Schnaiberg. 2008. *The Treadmill of Production.* Paradigm.

Haraway, D. J. 1991. *Simians, Cyborgs, and Women: The Reinvention of Nature.* Routledge.

Harding, S. 1991. *Whose Science? Whose Knowledge?* Cornell University Press.

Harding, S. 1998. *Is Science Multicultural?* Indiana University Press.

Huber, J. 2010. Upstreaming Environmental Action. In *The Ecological Modernisation Reader*, ed. A. Mol, G. Spaargaren and D. Sonnenfeld. Routledge.

Idle No More. 2014. Manifesto. http://www.idlenomore.ca/manifesto

Ihde, D. 1990. *Technology and the Lifeworld: From Garden to Earth*. Indiana University Press.

Illies, C., and A. Meijers. 2009. Artefacts without agency. *Monist* 92 (3): 420–440.

Jevons, W. S. 1906 [1865].*The Coal Question*. Macmillan.

Latour, B. 1993. *We Have Never Been Modern*. Harvard University Press.

Latour, B. 2009. A Collective of Humans and Nonhumans: Following Daedalus's Labyrinth. In *Readings in the Philosophy of Technology*, ed. D. Kaplan. Rowman and Littlefield.

Leopold, A. 1949. Sand County Almanac.

Lomborg, B. 2001. *The Skeptical Environmentalist*. Cambridge University Press.

Lomborg, B. 2008. *Cool It: The Skeptical Environmentalist's Guide to Global Warming*. Knopf.

Marx, K. 1991. *Capital*. vol. 3. Penguin.

Marx, K., and F. Engels. 1975. *Selected Correspondence*. Progress.

Mol, A. P. J. 1995. *The Refinement of Production*. Van Arkel.

Mol, A. P. J. 1997. Ecological Modernization. In *The International Handbook of Environmental Sociology*, ed. M. Redclift and G. Woodgate. Elgar.

Mol, A. P. J. 2001. *Globalization and Environmental Reform: The Ecological Modernization of the Global Economy*. MIT Press.

Mol, A. P. J. 2002. Ecological modernization and the global economy. *Global Environmental Politics* 2 (2): 92–115.

Mol, A. P. J., and M. Jänicke. 2010. The Origins and Theoretical Foundations of Ecological Modernisation Theory. In *The Ecological Modernisation Reader*, ed. A. Mol, G. Spaargaren and D. Sonnenfeld. Routledge.

Mol, A., G. Spaargaren, and D. Sonnenfeld, eds. 2010. *The Ecological Modernisation Reader*. Routledge.

Pellow, D. 2007. *Resisting Global Toxins*. MIT Press.

Polimeni, J., K. Mayumi, M. Giampietro, and B. Alcott. 2008. *The Jevons Paradox and the Myth of Resource Efficiency Improvements*. Earthscan.

Reo, N., and K. P. Whyte. 2012. Hunting and morality as elements of traditional ecological knowledge. *Human Ecology* 40 (1): 15–27. doi:10.1007/s10745-011-9448-1.

Schnaiberg, A. 1980. *The Environment*. Oxford University Press.

Schnaiberg, A., and K. A. Gould. 1994. *Environment and Society*. St. Martin's Press.

Simon, J. L. 1981. *The Ultimate Resource*. Princeton University Press.

Steingraber, S. 2010. *Living Downstream*. Da Capo.

Whyte, K. P. 2013. On the role of traditional ecological knowledge as a collaborative concept: A philosophical study. *Ecological Processes* 2 (1): 1–12.

Wood, D. 2001. What is ecophenomenology? *Research in Phenomenology* 31 (1): 78–95.

York, R., and B. Clark. 2010. Critical materialism: Science, technology, and environmental sustainability. *Sociological Inquiry* 80 (3): 475–499.

4 Resistance to Risky Technology: When Are Our Environmental Fears Justified?

Paul B. Thompson

We might as well start with Ned Ludd, the apocryphal figure put forward as the perpetrator of sabotage against the factory system in England during the early decades of the nineteenth century. The Luddites were engaged in a protest against the social effects of mechanization in the textile industry. Not that knitting and weaving weren't already technologically intensive activities at the time. The shift was from a household craft industry to more specialized machinery organized on the shop floor and powered by a central engine. The term "Luddite" has come to be equated with futile and irrational resistance to technical change, but the Luddites of nineteenth-century England were aware of the way that the new factory system was shifting the economic return on textile production from skilled household labor to capital. They did not like it.

The Luddites were not particularly concerned about the environmental effects of the industrial revolution. It was not really until 1864, when George Perkins Marsh published *Man and Nature*, that accounting for the environmental toll of technological advance began in earnest. (For a recent edition, see Marsh 2009.) In retrospect we can see the mining of iron and coke to build factories and the burning of coal to power them as an early warning of the hidden cost behind technologically based production efficiencies. Iron and steel works, Watt's stationary steam engine, and the subsequent development of machinery to reorganize the production process for virtually all types of manufacturing were an early form of "emerging technology"—a shift in technical practice applying recently discovered principles for manipulating matter, developing artifacts and re-organizing work processes. Like subsequent technological revolutions, the cluster of tools and techniques that made the early industrial revolution possible also precipitated social and environmental effects that were poorly understood at the time of their initial deployment. When in the rare case impact on air or water quality was recognized, environmental impact was largely ignored.

In the twenty-first century, opposition to novel types of technology shares several features with the protest of the Luddites. First and foremost, it is almost always mocked and derided among decision makers and technical elites as irrational and ill informed. Proponents of new energy technology, biotechnology or nanotechnology trumpet efficiencies and potential benefits, and while they acknowledge risks, they portray their critics as ignorant and fearful, if not backward looking and irresponsibly committed to old ways. However, many opponents of novel technology have a social agenda that is highly credible, at least on *prima facie* grounds—a second feature that latter-day critics have in common with the Luddites. Novel technologies inevitably displace older ones, and the workers, farmers, and craftspersons who derive a living from the older ways are rarely among the beneficiaries of emerging technology. New technologies that involve new materials also cause changes in land use, and the tenuous hold that marginalized groups and individuals have on their homelands can be the first casualty of technical change. Change in land use is the first link to environmental impact. While earlier generations of Luddites interpreted this impact in largely social terms, concerns about risk to health, to biodiversity, and to the integrity of habitats, ecosystems, and landscapes have now become incorporated into the litany of uncompensated harms that are routinely recited when new technologies are proposed.

The Basic Story Line for Novel Technology

It is important to recognize that there is always a rationale for technological innovation—a basic story line that should be viewed as the dominant social ethic for evaluating novel technology. The key assumption is that technological innovation is a key source—perhaps *the* key source—of greater efficiency in all manner of production processes. Efficiency is, in basic terms, a measure of how effectively a given set of tools and methods utilize raw materials and human labor in the creation of artifacts, valuable goods or other desired ends. A process that gets more of what people want from the same amount of effort and materials is more efficient. But production can also create *un*wanted outputs such as pollution. Accurate measures of efficiency reflect this by treating unwanted outputs as a cost of production. Producers may have little incentive to account for pollution costs when they are born by people distant in space or time. Efforts to develop more accurate accounting and effective incentives for pollution costs are the primary focus of environmental economics, but this will not be a primary focus in this chapter.

Efficiency in production processes is, in turn, thought to be one of the most reliable drivers for economic growth. Although the theoretical nexus of innovation, efficiency and economic growth emerged gradually in the 1600s, a particularly cogent and persuasive account of them can be found in *The Wealth of Nations* (1776). Adam Smith offered a memorable discussion of *division of labor* in which he describes how pin production can be organized. He convincingly shows that it would be an onerous task for a single person to smelt raw metal, draw it out, cool it and then polish and sharpen it to make a pin. But when these tasks are specialized, skills and tools can be refined. A properly equipped work team of a dozen can produce hundreds of pins in the time that each one of them a single pin, resulting in a productivity gain of an order of magnitude.

The process that Smith described for manufacturing can, in the view of many economists, be repeated in virtually any sector of the economy. Consider how the invention of recording technology or motion pictures allow the 'production' of a musical or theatrical performance to achieve unheard-of advances in efficiency. The artists themselves expend roughly the same labor that would be needed to entertain an audience of several hundred. The technological capture of the performance requires the additional labor of many individuals, it is true, but the result is that the performance can now be enjoyed by virtually limitless numbers of individuals. The twentieth century saw a virtual parade of technological advances, each of which has reduced the costs (increasing the efficiency) of the production process. The comparison between Edison's wax cylinders (made roughly twelve to twenty at a time) and a performance distributed digitally over the Internet to an unlimited number of tablet computers illustrates the efficiency gains. The fact that many more people are participating (as both producers and consumers) testifies to its contribution to economic growth.

There are flies in this ointment, of course. Many of them relate to the social consequences of productive efficiency and economic growth. Do Smith's factory workers derive any benefit of their cost-efficient approach to manufacturing pins, or are skilled artisans simply replaced by unskilled labor receiving the minimum wage? In general, does a disproportionate share of the increased efficiency from novel technology go to the people who own the machines when compared to workers on the shop floor? Perhaps much of the benefit goes to consumers in the form of lower prices, but if there is not *some* significant share of benefits going to the people who develop more efficient technology, the incentive for innovation evaporates altogether. And then what is the role for government in all this? Should government intervene in the private sector to protect workers from exploitation? Should government seed technological innovation through scientific research, and then speed it along by removing obstructions that may exist in the form of antiquated regulations or work rules?

The way that people see the answers to these questions in different terms is what separates the political left from the political right to a very large degree. Activists and politicos may disagree about how the benefits of innovation and economic growth should be distributed, and they may have different views on how government should be involved, but they all assume that the main thing that novel technology does is to increase the size of the pie. Given this assumption, the problem is never with new technology itself, but with the way social institutions and public policies shape its implementation.

Environment Enters the Basic Story Line

The basic story line seems to suggest that technological innovation and economic growth can simply continue indefinitely—but there are obvious reasons to question whether that is correct. Early hints of the limits to innovation and growth were sounded by Thomas Malthus, whose mathematical model of the relationship between human population growth and the technological efficiency of food production predicted a dismal future for humankind. More generally, economists recognized any form of production requires material inputs in the form of raw materials and energy to power the production process. Indeed, it rapidly became apparent that the increased efficiency of human labor was being achieved by using more raw materials and fuels. To the extent that the source of these inputs is finite, total production cannot simply grow *ad libitum*.

Serious economic analysis of when and how the consumption of natural resources would have limiting effects on industrial economies began in 1962 with a book by Harold Barnett and Chandler Morse entitled *Scarcity and Growth*. Barnett and Morse's study launched the field of environmental economics. In general, it is fair to say that economists who entered the field have had environmentalist tendencies. It is also fair to say that some basic principles of economic theory militate against the possibility that even environmentally inclined economists would turn against technological innovation. The authors' finding is that as a manufacturing or production process begins to exhaust any given resource, it will by definition become more and more scarce. Its price will therefore increase, giving innovators a strong incentive to begin economizing on the use of that resource. The next round of innovation will make more efficient use of scarce resources, or will eliminate the need for them entirely by finding a more plentiful substitute.

The more serious economic problem occurs when faulty policy or social institutions fail to provide an adequate incentive for conservation or for innovations that help economize the use of scarce natural resources. The problem here is closely related to the neglect of pollution costs. In a nutshell, it arises when someone can deplete a resource without having to bear the cost of having done so. Take biodiversity. The diversity of species and genotypes is widely recognized to be a source of potential discovery and innovation. But the individuals or the generation of people that cause harm seldom experience the harm from loss of biodiversity. From an economics perspective, the problem is one of getting a current user to recognize that cost and to reflect it in his or her decision making.

The upshot is that no one who has made an appearance in this story line—not the workers, not the consumers, not the environmentalists, not the politicos on the left *or* the right, and certainly the innovators—has a bad word to say about technological innovation as such. Our social institutions, on the other hand, come in for all manner of criticism for the way incentives for innovation are structured and the benefits of innovation are distributed, but innovation as such is unalterably a good thing.

Fast Forward to Emerging Technologies

The primary emerging technologies of the late twentieth century and early years of the twenty-first have all been imagined by scientists, engineers, government officials, and business entrepreneurs who were versed in the basic story line. Most truly novel forms of technology have had either their origins or substantial development in conjunction with war. The Manhattan Project is the signature episode. Although scarcely understood before World War II, atomic energy was an important source of both weaponry and peacetime uses by 1950. The development of nuclear reactors for submarines spawned research into their potential for cheap and clean generation of electrical power. Nuclear reactors were going to be so efficient at providing power to the residential grid that electricity would be too cheap to meter.

Hundreds of nuclear powered generating facilities were built worldwide, and some nations (notably France) rely on them for more than half of their total electrical supply. Nevertheless, nuclear power set a pattern for environmentally based resistance that has become familiar for a series of other innovative clusters of technology. Most especially, agricultural and food applications of recombinant DNA technology have foundered as a result of significant opposition on environmental grounds. At this writing, nanotechnology and synthetic biology wait on a precipice. It is not clear whether the widespread opposition that has retarded the development of biotechnology will overtake these new approaches. It is also not clear whether biotechnology can make a comeback through a series of "killer apps" that will play a role in meeting the energy or food needs of the coming decades. Notably, both the medical biotechnology and the information technologies that have transformed so many aspects of life seem to have been immune from skepticism and opposition.

There are two related philosophical questions to be raised by each of the aforementioned clusters of tools (artifacts and techniques). First, how can we *explain* the outrage and skepticism leveled against agricultural biotechnology, synthetic biology and nanotechnology, and the relative pass given to medical (or so-called "red") biotechnology and the computer/cell phone/tablet revolution? Second, and perhaps most important, what should we think of these technologies on environmental grounds?

The first thing to notice is that is virtually impossible to be "neutral" about emerging technologies. Given the momentum supplied by the basic story line of technological innovation, anything less than deep suspicion amounts to a kind of support for their development and implementation, at least under appropriate safeguards. A second point to observe is that if the answer to the second question is solidly negative—i.e., that these technologies should be opposed—then the outrage and the skepticism become relatively easy to explain, especially in light of the level of public and private investment in emerging technology and the seemingly inexorable march of these tools and techniques toward regulatory approval and implementation.

Take, for example, "synthetic biology," a fairly straightforward development of biotechnology. At this juncture in its development, synthetic biology promises to extend the capabilities of biotechnology through standardized biological parts and the synthesis of whole genomes. Although both of these developments are promising, they are not, at this writing, realized technologies. The goal of standardized biological parts is to create a catalog of gene sequences with known functions that can be reliably strung together to produce known effects. This would bring biotechnology a step closer to the "Lego" analogy, and would allow for true engineering design principles to be applied to organisms. Whole-genome synthesis is the ability to construct a functional molecule of DNA from chemical bases. Though conceptually simple, it continues to face significant hurdles before becoming a practical technology. Together, standardized biological parts and whole-genome synthesis would make it technically feasible to draw up genomes—biological blueprints—for organisms that have not evolved in nature.

Nanotechnology encompasses an even larger array of possible technological approaches and methods. The basic idea is to learn how to use physical and chemical capabilities of materials at the nanoscale. At this near-atomic scale, ordinary materials take on novel properties. Gold reflects red on the light spectrum. Carbon can be shaped into uniquely strong and durable shapes. Other metals become flexible—allowing for the shiny "metalized" lining inside bags of potato chips, for example. The electrical conductivity of some materials is magnified, allowing for even faster computers and more efficient electrical processing. Forces such as momentum from the spin of an electron can be mobilized to do work. In truth, many uses of the nanoscale have been in use for some time.

As a result, even a superficial overview of the tools and techniques falling under the heading of emerging technologies would exceed the patience of its readers. Let's just say that the number of specific tools and techniques is constantly increasing, and that the discussion of what these emerging technologies actually involve will be limited to a small number of illustrative, but nonetheless selected and therefore potentially misleading, examples.

What Makes Emerging Technology So Disturbing on Environmental Grounds?

The simple (and not entirely incorrect) answer to this question is that these technologies all involve very new ideas with which the human species has relatively little experience. The cumulative atmospheric effect of burning the coal and gas that fueled the industrial revolution provides an impressive object lesson for unanticipated environmental impact. The human and environmental effect of chemical pollution provides another. Such simple observations are often cited in support of the Precautionary Principle: Don't wait for proof of harm before taking action against environmental insults. This kind of reasoning may be enough to turn many people against emerging technologies, but there are two large philosophical problems associated with it.

First, it does not explain why information and medical technologies should get a pass. Our computers and cell phones are replete with rare earths and toxic chemicals. Their extraction and disposal are causing environmental problems. Medical residues and waste are also serious pollutants. Pharmaceuticals in the environment (PIEs) include antibiotics and antipsychotics that are having measurable effects in wildlife populations, and may well be contaminating water supplies used for food production and human use. The explanation here may be that people just see the benefits of computers and drugs *for them*, and they don't see personal benefit when it comes to more productive crops, much less nanotechnologies and synthetic biology products that are yet to be well defined. This may suffice as a psychological account, but it is not an acceptable philosophical reason for making the distinction between acceptable and unacceptable technology.

More significant, the Precautionary Principle focuses on acting before *proof* of harm, not before any *indication* of harm. Agricultural biotechnologies, in particular, must be evaluated in comparison to unsustainable chemically and energy intensive production practices that are used widely in industrial food systems. Bt crops for example, have been genetically engineered to produce a toxin (*Bacillus thuringiensis*) that helps plants resist damage from caterpillars. Relative to production systems that utilize chemical pesticides, the health impacts on field workers and on other insects (hence biodiversity) are clearly positive. This is not a definitive defense of these products, to be sure. There may be still better options. Yet if the consequence of declining Bt crops on precautionary grounds is the continued application of highly toxic broad-spectrum chemical pesticides, it is a result that defies the very logic of the Precautionary Principle.

No one denies that virtually all agricultural technologies involve some element of risk. Indeed, the basic premise of present-day risk assessment is that everything we do is statistically associated with the materialization of one hazard or another. The rationale behind the risk-based approach to the evaluation of technology is *relative risk*: Ethics requires us to weigh the risks and benefits of the alternatives we have and, other things equal, to choose technologies that pose relatively lower levels of risk. This is neither trivial nor simple, as there are multiple hazards (to workers, to consumers, to flora and fauna, to habitat, etc.) that must be taken into account. While agricultural biotechnologies may indeed pose some unknown risks, these risks should be weighed against the known risks we associate with our current methods of agricultural production. When we follow this procedure, the question "What makes emerging technology so disturbing?" becomes interesting and important on philosophical grounds.

The Risk-Assessment Response

If emerging technology is disturbing and precaution seems warranted, the classic rational response (as articulated by Bentham) is to make a rational evaluation what the risk is before making a decision. Environmental and food-safety risk assessments follow a procedure that can be broadly described as involving three interconnected activities: hazard identification, exposure assessment, and risk management.

Hazard identification is the attempt to ascertain what can possibly go wrong. For example, plant tissues—notably those of potatoes—contain toxins that are harmful to human health. These toxins are not normally found in the edible parts of food plants, but biotechnology *as well as* ordinary plant breeding can cause the appearance of a toxin in a previously edible variety. Inadvertent introduction of toxins is a standardly recognized hazard in both plant breeding and biotechnology. Other hazards include unwanted impacts on wild ecosystems and uncontrolled spread of cultivated varieties. Hazard identification is always an inductive procedure, but when done by competent biologists having a good knowledge of molecular, genotype, phenotype, and ecological interactions it can produce a remarkable list of things that might possibly go wrong.

Exposure assessment or quantification is a process of estimating how likely it is that any hazard will actually materialize, given a set of specified initial conditions. In environmental risk assessment it is done with both models and experiential data; in toxicology it can be done with laboratory tests. When combined, hazard identification and exposure assessment result in an estimate of risk. Scientists and policy makers schooled in these techniques hesitate to use the word "risk" in cases where hazards are merely speculative or to indicate broad uncertainties. From the standpoint of a biological risk assessment, risk is a function of both hazard and exposure. In more classic ethical terms, risk reflects both the harm or damage that might be done and the probability or likelihood that the damage will actually occur.

The final phase is *risk management,* which may or may not include risk communication. This is the point at which someone decides what to do about a risk: ban the product, monitor it, allow it to be used but compensate losers, develop insurance, tell people about it, or simply let it go unregulated. All these and more are options for deciding what to do about a risk.

Although in some obvious sense, risk management comes after the quantitative or scientific measurement of risk, these processes interact with one another. In regulatory agencies, risk management may be limited by the agency's legislative mandate. The US Department of Agriculture, for example, regulates biotechnology under a law that authorizes them to control plant pests. Hazards that have nothing to do with plant pests are unlikely to surface in a USDA risk assessment.

Risk assessors do the best they can given the knowledge they do have, although their ability to get more knowledge may be constrained by resources, time, and the current state of the science. What is more, risk management unabashedly involves judgment as to which interests are most relevant, and what means are appropriate to protect them. It is, in one sense, a straightforward example of the utilitarian advice to follow the greater propensity of benefit and harm. It is especially compelling in those cases where it is impossible to avoid all possibility of harm whatsoever. Those who advocate risk assessment see it as the ethically correct response to the uncertainties that generate a call for precaution. It does *not* produce certainty, but neither does it require "proof of harm" (Shrader-Frechette 1980, 1983, 1991; Cranor 1990, 1993, 1997, 1999; Hansson 1989).

What risk assessment *does* accomplish is a scientifically informed estimate of the biological risks posed by a novel technology. It yields something more than a guess—but something less than a certainty—of how likely it is that health and environmental hazards will actually materialize. It then applies whatever ethical criteria have been identified for risk management to this estimate to yield a recommendation on what should be done. Risk assessment is not *contrary* to the precautionary stance that arises in light of the disturbing possibilities of emerging technology; indeed the argument for a precautionary approach and the argument for risk assessment are one and the same. But while a purely precautionary standard might be quite comfortable with remaining totally ignorant about the potentially disturbing things that technology might cause, risk assessment presumes that one is open to the possibility that some risks will be taken, while others will be rejected.

And there is one more point to emphasize before moving on. A scientifically informed hazard identification process can lead to the recognition of potentially harmful outcomes that someone working from pure intuition might never suspect. Sometimes these hazards are caused by the novel technology, but sometimes they are lurking in the status quo—the things we have and are *already* doing. One problem with the basic logic of precaution is that in being concerned about the risks of something new, one may have a disturbing tendency to underestimate the risks of something old. Risk assessments are thus not necessarily geared to prevent or forestall the emergence of novel technology, though they are designed to anticipate the biological outcomes that, in some cases, make us decide that the new new thing is not such a good idea after all.

The Social Amplification of Risk

Working with Ortwin Renn, Roger and Jeanne Kasperson developed the idea of "social amplification of risk" as a way to study and understand an array of socially based phenomena that account for deviations from the rational approach implied by biological risk assessment. In some cases, these deviations appear to be ethically unsupportable; in other cases they provide good reason to question whether the estimate of risk derived through hazard identification and exposure quantification is ethically adequate. The Kaspersons were originally interested in phenomena or features of a social situation that make the risks of an emerging technology greater than the biological experts tended to suggest; or alternatively, made the risks *seem* greater, causing social resistance and a strong divergence of opinion about the acceptability of risks. While social phenomena can sometimes cause people to see a situation a *more* risky than biological experts predict, they can also lead people to under-appreciate or be dismissive about risks. The general framework of risk amplification provides a philosophically fecund way to approach emerging technology, though in fact it is calling attention to the way that social and perceptual factors mediate the nature and level of risk, sometimes with amplifying effects and sometimes with dampening effects (Kasperson et al. 1988; Kasperson and Kasperson 1992).

Two of the most basic aspects of social amplification should be utterly uncontroversial from an ethical standpoint. First, biological risk assessment attempts to estimate hazards that affect life processes and living things, either the life of individual organisms (especially human beings) or populations and ecosystems. It is not designed to address many potential impacts that people care about. The list is long, including historic preservation; community development or integrity; jobs and employment; the viability of a business, a school, or a community organization; and the fate of a neighborhood, a town, or a tradition. These factors are simply left out of a biological risk assessment, largely because the discipline of biology does not have the concepts necessary to understand them.

A second framework for evaluating risk is that of informed consent. While there may be cases where we think that the appropriate way to evaluate a project is to compare risks and benefits, in other cases we think that the responsible course of action is to put the person or people who will bear a risk into a position where they can decide for themselves whether or not to do so. It should not be surprising that people who feel that risks are being imposed on them will object, calling for policy responsibilities (such as labels) that will put them in a position to decide for themselves whether or not to accept the risk of a new technology.

In the first case, the potential for unwanted or harmful consequences accruing from new technology is greater than what was anticipated in the typical risk assessment because the typical risk assessment deliberately omits certain types of harm. Even if a risk assessment is fully adequate in its anticipation of *biological* hazards, risk assessments performed for regulatory purposes do not even attempt to quantify hazards to a person or group's social or economic well-being. This is the kind of risk amplification that the Luddites would have understood implicitly. In the second case, it is less clear that a quantitative increase in risk is taking place. What may be happening here is that opportunities to accept or reject a risk are being denied. It is the seriousness or political contentiousness of the risk that is being increased or amplified. The risk is accompanied by a social phenomenon that has been characterized in the literature of risk studies (e.g., Sandman 1989) as "outrage." When risk creates outrage, opposition and resistance to the technological change are amplified.

The social amplification of risk framework should thus be seen as a general framework for understanding how biological risks fail to capture all of the factors that are ethically or politically relevant for social decision making on novel technology. This framework is not solely or even primarily oriented toward the classic questions of environmental philosophy, as the Luddite opposition to technological innovation on socioeconomic grounds attests. Nevertheless, environmental impacts have proved to be among the most prominent and politically contentious hazards during the past 50 years. Hence, social amplification of risk is an appropriate though sometimes neglected topic for environmental philosophy.

Social Amplification: Cognitive Heuristics and Biases

Research in psychology and research in behavioral economics have identified persistent patterns of deviation from putatively rational norms in human decision making under conditions of uncertainty and risk. Here we will note only a few of the ways in which decision appears to be biased.

Availability People will make estimates about the likelihood or the statistical distribution of events based on information that is familiar to them, even if they have little reason to presume that information relevant. People who know someone who has had a heart attack recently will persistently estimate the frequency of heart attacks throughout the population much higher than those who do not.

Aversion to losses People tend to evaluate two quantitatively identical probabilistic decisions differently when they are framed in terms of gains or losses. People who will reject a bet when told that they have a 60 percent chance of losing will accept the bet when told that they have a 40 percent chance of winning. Standard conceptions of rationality find these betting preferences to be logically inconsistent.

Anchoring Faced with a choice with uncertain outcomes, the first information people are given will influence their estimate of likelihood, even when they have good reason to think that the information is meaningless. Amos Tversky and Daniel Kahnemann conducted an experiment in which they spun a roulette wheel and then asked subjects whether the resulting number accurately predicted the number of African nations in the UN. People speculated that the correct number was significantly higher when the random number was high than when the number was low. A rational decision maker would not be influenced by a result known to be both random and unrelated to the question at hand.

Adjustment bias A related phenomenon occurs when people resist proper adjustment of probability estimates in the face of new information. People observing black and white balls drawn from a bag will resist appropriate revision of an initial random guess of the ratio between white and black balls.

These results suggest several possible implications for social resistance to emerging technologies. First, they provide a basis for explaining away public fears. If concern about the risks of nuclear power, biotechnology, or nanotechnology is simply a result of the way in which information about these technologies has become available, psychological deviation from "rational" behavior is understandable. Given that speculative environmental and health impacts can almost always be associated with new technology, there are ample opportunities for aversion to losses and anchoring to set in. Experience with pesticides, with the Chernobyl accident, or with mad cow disease may make adverse judgments especially available to people. Once established, adjustment bias would give these adverse judgments enduring influence. These psychological biases thus contribute to the social amplification of risk, even though there is no valid normative basis for taking them seriously. Rational decision making would dictate that public fears be ignored or discounted by policy makers, and that objective, scientifically based biological risk assessments be used as a basis for political decision making regarding novel technology.

However, there may another sense in which these heuristic devices for dealing with uncertainties and incomplete information *are* rational. After all, if they were truly and consistently leading human beings to make suboptimal decisions, evolution should have selected against such a cognitive architecture. Thus, there may be social or evolutionary benefits to these risk heuristics that are not captured by conventional conceptions of rationality. They may, for example, facilitate more rapid decision making and social cohesion while yielding decisions that satisfy with respect to outcomes under the uncertain conditions that humanity has faced over evolutionary history. They may economize on decision-making resources that are not taken into account in traditional conceptions of rational choice.

Whether or not the deviations from rational optimization that are associated with cognitive bias are regarded as departures from rationality, the fact that they appear to part of human cognitive architecture certainly implies that people who are influenced by them are behaving very much as we would expect people to behave. There is a problem for the view that they should simply be ignored by policy makers, especially in democratic societies. We should expect that decisions running counter to the widespread risk preferences of people influenced by cognitive bias will be unpopular, and that it may spark social movements to reverse them. To the extent that decision makers are expected to reflect the preferences of citizens, it may be appropriate for them to make decisions that are consistent with cognitive bias, rather than what theoretical conceptions of rational decision making dictate.

Socially Amplified Risk: Rational or Irrational?

In summary, a long list of independent factors can amplify or seemingly increase the risks associated with emerging technologies. But are these amplifications rational? Do they distort the "real" risk, or do they reflect the reality of life as lived more faithfully than the idealized quantifications produced by scientific methods? Elsewhere I have characterized the contrasting philosophical responses in terms of purification and in terms of hybridization (Thompson 1997, 2012). Although advocates of purification can acknowledge the legitimacy of social costs, they resist any tendency to permit the outrage and distrust that might felt by persons or groups who fear social consequences to license an amplification of risks that are regulated on grounds of hazard to environment or human health. Since statutes in industrialized countries—especially the United States—typically limit the authority of regulatory agencies to consider socioeconomic consequences (such as job loss, or a decline in property values) in their decision making, advocates of purification will insist that risk-based decision making in these agencies be shielded from political pressure that arises in conjunction with socioeconomic threats and the public outrage that they can cause (Sunstein 2004, 2005).

The hybridizing perspective might hold that risk is, above all else, a classification or a category that functions to highlight and motivate responsive action. Those factors that contribute to distrust represent wisdom gained from experience with chemical pesticides, with industrial air and water pollution, and with the deceptions of wealthy and well-placed actors who have persistently benefited at the expense of the less powerful and less fortunate majority. To insist on careful parsing of biological and social hazards is, on this view, to ignore these lessons and to invite further exploitation. In comparison to the accumulated harm done by the powerful and irresponsible actors in our society, there is little if any harm to fear when victims of oppression gain a victory because health or environmental fears helped mobilize widespread resistance and opposition. And the reluctance of powerful actors to permit labeling of emerging technologies only proves further that their motives are not to be trusted. Sheldon Krimsky may offer the best example of a hybridizing perspective on risks among present-day environmental philosophers (Krimsky 2007; Krimsky and Wrubel 1996).

The view that I have advocated attempts to find the narrow window of opportunity between the purifying and the hybridizing perspectives. It acknowledges the epistemological appeal of purification while recognizing the compelling political appeal of hybridization. I hold that if one is to maintain the categorical distinctions implied by distinguishing biological and social hazards and the norms of practice associated with risk quantification based on the best scientific evidence, one must also accept a moral obligation to support ethical norms and political institutions that allow socioeconomic outcomes and inequalities of all kinds (including unequal access to information) to be addressed—in other words, no "scientific risk assessment" without effective responses to injustice on the ground.

Allowing epistemic norms and carefully developed conceptual categories to reinforce and perpetuate social injustice is ethically unacceptable (Thompson 1997, 2012). This view accepts the logical and epistemological orientation of purification on the grounds that norms of scientific practice are necessary for environmentally sound judgment, but rejects the way that these norms have been allied with powerful social interests that have allowed environmental *and* social injustices to persist. Importantly, I claim that scientific practitioners cannot simply leave the ethical work to others. The practice of environmental science *demands* a response to social injustice.

In closing, I would note two reasons to doubt the relationship between science and injustice. First, the point of the social amplification literature was to explore psychologically and institutionally based sources of divergence in expert and non-expert opinion. As was shown above, there are situations in which the expert gets it wrong and situations in which the non-expert gets it wrong. An approach that privileges non-expert perspectives over those of experts simply ignores the underlying issues that I have been attempting to present in this chapter, rather than resolving them. Second, non-experts may be as likely to *underestimate* the risks they bear through institutions that create systematic forms of information inequality. That they find the risks of emerging technologies "acceptable" may simply reflect a form of ignorance for which the expert should actually be held culpable.

References

Cranor, Carl F. 1990. Some moral issues in risk assessment. *Ethics* 101 (1): 123–143.

Cranor, Carl F. 1993. *Regulating Toxic Substances: A Philosophy of Science and the Law*. Oxford University Press.

Cranor, Carl F. 1997. The normative nature of risk assessment: Features and possibilities. *Risk* 8: 123.

Cranor, Carl F. 1999. Asymmetric information, the precautionary principle, and burdens of proof. In *Protecting Public Health and the Environment: Implementing the Precautionary Principle*, ed. C. Raffensperger and J. Tickner. Island.

Hansson, Sven Ove. 1989. Dimensions of risk. *Risk Analysis* 9 (1): 107–112.

Kasperson, Roger E., and Jeanne X. Kasperson. 1992. Determining the acceptability of risk: Ethical and policy issues. In *Assessment and Perception of Risk to Human Health*, ed. J. T. Rogers and D. V. Bates. Royal Society of Canada.

Kasperson, Roger E., Ortwin Renn, Paul Slovic, Halina S. Brown, Jacque Emel, Robert Goble, Jeanne X. Kasperson, and Samuel Ratick. 1988. The social amplification of risk: A conceptual framework. *Risk Analysis* 8 (2): 177–187.

Krimsky, Sheldon. 2007. Risk communication in the Internet age: The rise of disorganized skepticism. *Environmental Hazards* 7 (2): 157–164.

Krimsky, Sheldon, and Roger Paul Wrubel. 1996. *Agricultural Biotechnology and the Environment: Science, Policy, and Social Issues*. University of Illinois Press.

Marsh, George Perkins. 2009. *Man and Nature: Or, Physical Geography As Modified By Human Action*. Harvard University Press.

Sandman, Peter M. 1989. Hazard versus outrage in the public perception of risk. *Contemporary Issues in Risk Analysis* 4: 45–49.

Shrader-Frechette, Kristin S. 1980. Technology assessment as applied philosophy of science. *Science, Technology & Human Values* 5 (4): 33–50.

Shrader-Frechette, Kristin Sharon. 1983. *Nuclear Power and Public Policy: The Social And Ethical Problems of Fission Technology*. Reidel.

Shrader-Frechette, Kristin S. 1991. *Risk and Rationality: Philosophical Foundations for Populist Reforms*. University of California Press.

Shrader-Frechette, Kristin Sharon. 2002. *Environmental Justice: Creating Equality, Reclaiming Democracy*. Oxford University Press.

Sunstein, Cass R. 2004. *Risk and Reason: Safety, Law, and the Environment*. Cambridge University Press.

Sunstein, Cass R. 2005. *Laws of Fear: Beyond the Precautionary Principle*. Cambridge University Press.

Thompson, Paul B. 1997. Science policy and moral purity: The case of animal biotechnology. *Agriculture and Human Values* 14 (1): 11–27.

Thompson, Paul B. 2012. Synthetic biology needs a synthetic bioethics. *Ethics, Policy & Environment* 15 (1): 1–20.

5 Getting the Bad Out: Remediation Technologies and Respect for Others

Benjamin Hale

Arguments for and against environmental remediation have tended to emphasize mitigation of harms while turning a blind eye to other moral considerations that inform our views on environmental wrongdoing (Nelson 2008; Singer 2006). In this chapter, I focus the discussion much more narrowly. I inquire into the conditions that make some very narrow set of mitigation projects permissible, and seek to outline what those conditions might be. Ultimately, I aim at the conclusion that what makes an engineering project permissible is whether all affected parties can accept not only the side effects of the project but also the legitimacy of the project itself. The problem for this chapter should be contextualized as part of a much larger set of questions oriented around addressing concerns in mitigation of climate change and in environmental remediation.

The extent to which affected parties should have a say in the permissibility, legitimacy, or justification of a proposed project may not appear immediately relevant to the question of how to proceed in the face of anthropogenic environmental messes. That is, one might think that if one creates a mess, it's just natural that one should clean it up. Certainly, if it's true that anthropogenic climate change is the result of our careless emissions, then we should do what we can to reduce the effects of our actions on the climate. That seems to be just a straightforward fact about moral messes.

In earlier work, I've tried to demonstrate at least three related points. First, I've argued that environmental problems go beyond simple characterization as damage done from harms. This is particularly true for ambient pollutants, of which greenhouse gasses fall into the paradigmatic category. Second, I've argued (with my colleague Bill Grundy) that remediation technologies (RTs) don't offer a simple fix to non-harm-related environmental problems (Hale and Grundy 2009). Having created a mess does not straightforwardly translate into a responsibility to clean that mess up. Nor does it even translate directly into permission to clean it up, or into permission to clean up the messes of others. Third, I've argued (with my colleague Lisa Dilling) that very-large-scale geo-engineering projects, such as ocean fertilization, are impermissible by virtue of the extent to which they are undertaken for the wrong reasons and the extent to which they are caught up in the lives of others (Hale and Dilling 2011). Dilling and I argued that actions such as ocean fertilization cannot be understood independently of the antecedent events that have coalesced to bring about their consideration as viable options.

In this chapter, I'll pick up where that work left off, with the central objective of identifying a class of permissible remediation technologies. This will require a relatively important preliminary observation: Not all RTs are created equal. Some, such as *in situ* bioremediation, involve modifying a specific site by introducing biological material to digest spills (Suthersan and Payne 2004; Warner 2007). Others, such as carbon sequestration, involve planting acres of new flora, or simply managing forests, in order to absorb carbon from the atmosphere (Dilling 2007; Dilling et al. 2003; Potter et al. 2008; Thom et al. 2002). Still others, such as ocean fertilization, seek to capture carbon by encouraging phytoplankton growth in the ocean, thereby manufacturing a mid-ocean red tide and sequestering carbon to the ocean floor (Buessler et al. 2008; Caldeira and Wickett 2005; Chisholm et al. 2001; Department of Energy 2008; Jamieson 1996; Kintisch 2007; Powell 2007; Scott 2005; St. Clair 1999). Yet a fourth technique, sometimes called "atmospheric scrubbing" or "air capture," involves capturing ambient air and scrubbing carbon or other pollutants from it (Herzog 2003; Jones 2009; Keith et al. 2006; Parson 2006; Spreng et al. 2007; Stolaroff 2006). Many of these technologies can be demonstrated to work, and in some cases to work very well. Nevertheless, each poses moderately to expansively thorny ethical challenges. Ethicists must be engaged not only in assessing the various implications of such technologies but also in establishing which RTs are appropriate and which are not. We should be careful to distinguish the question of *whether* remediation is permissible from the question of *which* remediation technologies are permissible.

My argument proceeds in four stages. In the first stage, I introduce a complicated scenario— the "Town of Incenter"—involving three companies and their emissions. I use this scenario to challenge intuitions about what the wrong of pollution consists in. In the second stage, I suggest that all remediation actions must be evaluated not just in terms of their consequences but also in terms of the reasons that most appropriately describe them. In the Town of Incenter case, this involves assessing not only the motivation for the action but also the antecedent conditions that have coalesced to bring about consideration of remediation in the first place. In the third stage of this argument, I introduce several situations related but similar to the Town of Incenter. I ply my position primarily along concerns about the direction and dissipation of agency, suggesting that actions for which we are responsible are actions over which we maintain some manner of moral jurisdiction. This position stands in contrast to a position that seeks to make a similar claim based on a false distinction between the natural and the non-natural, or the anthropogenic and the non-anthropogenic. I do not cover the question of moral jurisdiction between single agents and collective agents, between the 'I' and the 'We', though this concern is circulating in the background. In the final stage I argue that the appropriate intervention comes only through technologies that shift the world back from state Y to state X, and not on to a third state, Z. I also suggest that there is some mitigating factor—here framed as the intersection of interests—that suggests that it isn't immediately permissible to bring the world all the way back to state X. As a result, I reason that technologies like, but not limited to, atmospheric carbon capture are easier to justify than many other proposed remediation technologies.

One quick and final observation on methodology: I am employing the much-maligned device of intuition pumping as my primary theoretical engine (Dennett 1995; Sencerz 1986). Naturally, some philosophers are skeptical of the usefulness of intuition pumps, as they are effectively a string of hypothetical thought experiments aimed at priming the reader's intuitions about a given set of problems. In many philosophical circles they are not employed at all. They are, however, employed to great effect by philosophers as diverse as Judith Jarvis Thomson and John Searle. I am well aware of the limitations of this device, but I think it particularly useful for assessing the sorts of problems that I will be addressing in this chapter. The reason for this is primarily pragmatic: Principle-oriented or value-oriented methodologies, common in many branches of philosophy, inevitably raise many more questions than they answer. Climate issues are exceptionally broad-reaching, touching on science, policy, engineering, business, and so on. The intuition pump has the benefit of starting from a point of intuitive convergence and working out from there (McMahan 2000). Though it may appear that this makes for a somewhat aimless stroll, I use the device only to inspire in the reader a reflective equilibrium, as many others have done before me (for examples, see Boonin 2003 and Rawls 1951). Rest assured,

I have a destination. The intuition pumps serve to bring the reader most efficiently to the position that I advocate.

Collective Wrongdoing

Suppose three widget companies: Acme, Beatme, and Capme. Acme emits additive A, Beatme additive B, and Capme additive C. Alone, each additive is completely undetectable and harmless to humans. When combined, however, the compound ABC is noxious and harmful to humans. Moreover, any non-ABC combination of two additives is undetectable and harmless to humans. AC, BC, and AB are all innocuous compounds. If Acme and Beatme were to continue production of their widgets in the absence of Capme, there would be no noticeable outcomes. So too if only Acme and Capme were operating alone, and so also if Beatme and Capme were operating in the absence of Acme. It is the *confluence* of the additives, in other words, that creates the negative outcome. To foreshadow, I am expressly avoiding actual real-world chemical combinations on the assumption that A, B, and C are variables that may be quite distinct, or may, in fact, stand for the same pollutant. They could all be carbon, for instance.

To avoid responsibility-related complications about first priority rights, suppose that Acme, Beatme, and Capme become operative at exactly the same time—January 1, 2009—and that they are distributed geographically at the axes of an equilateral triangle. Depending on how the winds blow, the town of Incenter receives a greater or a lesser degree of compound ABC. Sometimes the winds will blow just so that A mixes with B and C on the east side of town, and sometimes the winds will blow just so that B mixes with A and C on the south side of town. Sometimes the winds will keep A, B, and C from mixing at all.

The first most obvious question, of course, is this: Which company, if any, is committing a wrong? Without clear harms from any single company, it may appear either that no company is doing anything wrong or that all companies are equally complicit in creating a collective wrong. Certainly there are many directions from which one could weigh in on this issue. As I've said, I'm not interested in these responses. I'm talking about distributed responsibilities here—circumstances in which multiple parties can be said to be co-responsible for having brought about a given state of affairs—and I want to know what we are entitled to do in the face of bad states of affairs that are brought about through the distributed actions of several actors.

Backward-looking responsibility questions such as those above are important, of course. They may ultimately be relevant to a determination of what to do. If we can identify a culprit, we may be able to force that culprit to take action while keeping everything else running smoothly. On the other hand, there may not be enough time. Perhaps, in view of the enormity of the problem, it is better to brush most of these concerns about culpability to the side and ask how to move forward. We can worry about culpability once the dust has settled.

This forward-looking approach is also tempting. To mitigate harms, we should stop any one of the three companies from emitting either A, B, or C. Since the origins of ABC are known, it is also known that reducing one of the three emissions will result in overall benefits. Unfortunately, there are many forward-looking policy approaches that can accomplish this, and each is beset with its own problems. We might simply shut one of the plants down, but to do so we would have to have a good method for determining which one to target. We might flip a coin and simply *destroy* one of the plants—just blow it up—a solution that might be effective, and perhaps even gratifying, but is arguably gratuitous and unnecessary. We might force a negotiation between the companies, so that two of the companies buy out the third company—an option that may be efficient but isn't necessarily optimal. We might regulate the pollutants A, B, and C, so that the townspeople of Incenter only receive smaller and less damaging amounts of ABC. We might control for discharges. And so on and so on. We could go on for a long time. Without some sense of the constraints under which we must operate, everything is on the table. This points to, among other things, a requirement to take a closer look at each of the three companies in order to determine what they're up to. I'll come back to this, because it is the question about the justification of the production project in the first place that, I think, will be driving the overall assessment of whether and how to proceed. But again, as with assigning blame, neither am I interested in the technical nor the policy solution to this problem. Rather, I'm interested in the morally appropriate remediation solution that brings harms back down to a tolerable level.[1] This is clearly the bigger and more pressing challenge—not how to bring the harms down, but what is permissible. One can easily propose a range of solutions to the problem, any one of which may trample some set of rights or principles that are of moral significance.

What then are the conditions under which it would be permissible to reduce the problem generated by the distributed actions of Acme, Beatme, or Capme?

One important observation is that remediation technologies ostensibly offer a middle moral ground: an opportunity to avoid assigning blame *and* an opportunity to avoid draconian policies of the sort I mention above. They are classic "win-win" situations that entice policy makers with an appealing third way. It is therefore mighty difficult to argue against RTs. Nevertheless, this hasn't stopped people from trying. Objections to RTs generally function by demonstrating that they are not, in fact, win-win situations, but that there is a cost or a loss somewhere (see, for example, Enkvist et al. 2007). In so doing, they tend to disregard concerns about the antecedent reasons and obligations to others, presupposing that the reasons describing an action take a back seat to other benefit and cost considerations. The most forceful objection that foes of RTs seem to muster is that RTs change our motivations (and, ergo, our "reasons"), thereby somehow encouraging bad behavior.

It is my view that the bigger danger is not that RTs change our motivations (and thus our behavior), but that they potentially mask what is morally suspect about our actions in nature in the first place: that our current practices are unjustified. The important question that the culpability responses I raised earlier elide, but that these policy responses underscore, is the question about who is doing what for which reasons. And it is that direction that I shall now pursue. I want to look closely at the reasons we have for undertaking action.

The Direction of Agency

One common reasons-related objection to remediation technologies is that they encourage bad behavior. As I mentioned earlier, there are many arguments for and against RTs, but here I am distilling out only *reasons-related* objections, by which I mean those that object to such technologies on grounds that they entail acting for the wrong reasons. The argument that RTs encourage bad behavior (or install a "moral hazard") is, to my knowledge, one of the few publicly articulated reasons related objections, as almost all other objections suggest that some particular method of remediation is too risky (Schneider 2006) or at best that that method of remediation may unjustly affect the lives of others (Jamieson 1996). I think the moral hazard is a wrong-headed concern, and I have argued against it elsewhere (Hale 2009). Nevertheless, it is worth exploring these cases, if only because it is a speedy route to the governing observation that harms from actions do not necessarily establish those actions as wrong, and that harms from wrong actions do not immediately authorize the source of that harm (or their proxy) to redress that harm.

What would it mean for me to act under the supposition that the world would be cleaned up immediately following my act? In other words, it is conceivable that the introduction of a cleanup technology makes it possible to reorder my priorities. In the face of RTs, I now have a method of acting without facing any repercussions from my actions. As a 2002 press release from the Los Alamos National Laboratory notably put it: "Imagine no restrictions on fossil-fuel usage and no global warming!" (Rickman 2002) Perhaps there is something wrong with thinking this way. Consider that airlines pay cleaning crews to remove garbage after patrons have disembarked from their flights, and that stadiums pay janitors to clean up after their games. As a patron with knowledge of this, am I permitted to leave my garbage behind on my seat? In one sense, yes, because the cleanup person is there to clean. There may even be a reasonable expectation from all parties that this is the way things are done. The airline has an interest in seeing its patrons disembark as quickly as possible, for instance, and so may not want them to bother tidying their waste. In another sense, it is clearly not permissible for me to leave my garbage on my seat. It's plainly inconsiderate of others to leave a mess behind. It's inconsiderate of those who will later sit in the seat, but more than this, it makes extra work for the cleanup staff. One could base arguments on a good number of external considerations about what would make the act permissible or impermissible. Perhaps the patron and the janitor are bound by a tacit contract, or by conventions, to abide by general rules of airline decorum. Or perhaps the contract is more explicit: Perhaps the janitor has been hired expressly to serve the patrons. We can go quite a distance by exploring the branches of this axis.

Putting such arguments aside, there is a point at which most agree that some actions overstep cleanup capacity and become clearly impermissible. It is not permissible for me to smear mustard on the seats, for instance. Despite the presence of the cleanup crew, I may be doing permanent damage to the seats if I smear mustard on them. Smearing mustard goes beyond any reasonable expectation of what the crew might be in place to do. It would obviously be wrong of me to deliberately smear mustard on the seat. If I were *accidentally* to smear mustard on the seat, I may not have done wrong, but I may be liable for having ruined the seats. The situation varies, of course, but it varies according to degree and scope. One very natural inclination when talking about RTs is to speak in terms of intentional or deliberate actions versus accidental actions. This view is handily covered in the literature on doing and allowing. (For a particularly poignant exploration of the role of intention regarding climate change and geoengineering, see Jamieson 1996.)

It is also natural to assume that what really should regulate one's behavior is whether harm is done. But again, sometimes we put these cleanup crews in place specifically because we want to encourage harmful behaviors. We tear holes in the walls of our home in order to get at the plumbing, knowing that we can repair the damage later. We conduct surgery on willing patients, knowing that we can sew them back up. We employ orderlies in hospitals so that doctors can focus on their patients and needn't worry as much about their messes, sometimes even with the expectation that sheets and linens will be damaged. We pay landowners large sums to permanently sully their land so that we can generate garbage without having to live in squalor. Harms are part of the figuration in such cases. And yet we very often say, in all of these cases, that some actions are permissible, that some are impermissible, and that there are still wrongs to be done.

What this points to, I think, is the importance of looking at the breadth of the act—of finding the most accurate act description and assessing the obligations and permissions according to that description. We should ask: What's really going on? How did such a state of affairs come to be? What reasons could be motivating one actor to take action? What reasons might be festering under the surface? Who will be affected or has been affected, and what stake can they claim in the consequent *re*-action? In order to determine what actions are permissible, we need to ask what reasons are justified, which involves taking into consideration all of these commingled reasons. All such related considerations—about the initiating wrong, about the intent of the actor, about the parties affected, and so on—are extremely important to a determination of the permissibility, or justifiability, of an action.

This discussion is revealing. In identifying permissibility as functioning according to the justifiability of an action, some basic principles begin to emerge about the constraints to which remediating actions should be subject: The more narrowly an action impacts the world, the fewer interests are involved, and the more latitude an actor has to alter the outcomes of the action.

In the case of emissions into ambient air, the permissibility of an action is immediately complicated by the extent to which the outcomes of the preliminary action pervade the lives and interests of others. Allow me a few more examples. Here are two intuitions I think we probably share, all related to somewhat more containable messes:

• It is permissible for me to camp at a campsite, so long as I can return the campsite to its original state upon my departure.
• Everything else being equal, it is not permissible for me to camp at a campsite with the intention of returning the campsite to its original state at a later time.

I cannot leave the area for a few days and then come back to clean up my mess. The reason for this is that others probably will, or simply just may, be affected by my actions in the interim. Hikers may stumble on my abandoned site and have their experience in nature ruined. Bears may find my site and develop a taste for peanut butter, thereby jeopardizing their taste for natural foods such as berries and salmon. Rodents may find temporary refuge in my abandoned cook tins. And so on. As the days tick by, the wrongness of my refusal to clean up the site becomes cemented. As my earlier action (the sullying of the campground) begins to overlap with the interests and concerns of others, it takes on the character of having disrespected them. My culpability for having done wrong becomes ossified in the moral fabric of interaction.

Here's a further thought: It is not permissible for me to leave a site unattended for 50 years and then come back and remove something that has been a part of that site with the same moral authority as I might under conditions in which I clean up the site on the day I leave, or even a few days later.

It is not clear that I am permitted to disturb mining equipment from the 1940s, for instance, even if I or my company left it there. In the intervening years, the abandoned equipment may have taken on a different importance or meaning. (It's not that it definitely *has* taken on a different meaning, only that it may have.) Hikers walking through the woods 50 years from now may stumble upon an archeological wonder. Future bears may have colonized the site to make it their own. Rodents may depend on the area for shelter. Trees and plants may have found the area particularly hospitable.

Just as agency emanates from an individual actor and offers up actions that can be deemed permissible or impermissible according to how they interweave with the interests of others, agency dissipates as time passes, as the outcomes of one past action intermingle with the lives and interests of others. This is no spooky metaphysical claim about agency, but rather a pragmatic reality. Singular agency dissipates: New considerations are born as lives and activities interact. From this, new reasons emerge. Minimally, time and historical considerations are introduced. The abandoned mining site becomes a place of historical significance.

Intuition Pumps

Consider the following cases.

Volcano

Suppose that there is some natural point source—a volcano, say—that in a steady state is emitting B into the atmosphere. Imagine that Beatme is taken out of production, but that this volcano emits equivalent quantities of B into the atmosphere. For clarity's sake, distinguish between B_V (emitted from the volcano) and B_B (emitted from Beatme). Now our scenario involves two online factories, Acme and Capme, as well as one volcano.

Is it permissible to remove B_V from the atmosphere in order to avoid its mixture into the harmful compound ABC? I think the answer is No. We ought not to remove B_V from the atmosphere, even though it is generating effectively the same emissions as the Beatme factory was emitting, and even though its emissions are commingling with A and C to create the toxic pollutant ABC, thereby creating the same harms on the population of Incenter. For reasons that I cannot cover here, I suspect my intuition would not waver in the face of reasonably inexpensive capture technologies that could be affixed to the mouth of the volcano. Despite my intuitions about atmospheric emissions of B_V, I also have intuitions that it may nevertheless be permissible to stop the volcano from erupting and covering the town of Incenter in magma, should such a technology be available. What I suspect is that the distinction between such cases hinges not on harms to individuals, nor on justice between individuals, but rather on *what could or would be accepted by all affected parties*. Being covered in magma is an unacceptable outcome to all. Filtering the atmospheric emissions of the volcano is a more questionable proposition, in terms of what could be accepted by all. We are thus left to find a different solution. In lieu of capping the volcano, we must seek to modify some output of Acme's or Capme's in order to alleviate the threat. As we interrogate this case, we should bear in mind the following somewhat more realistic case.

Suffusion

B is a man-made pollutant emitted from Beatme (in the form of B_B), but it also is emitted naturally, as from a nearby volcano (B_V), thus creating problems for Acme and Capme.

Would we be authorized in removing *more* B from the atmosphere than Beatme has emitted (B_B) to reduce the effects of ABC? I think the answer is No, that we wouldn't be permitted to remove more B from the atmosphere. We cannot remove B_V, I believe, but can remove only as much and up to B_B, the amount that Beatme has contributed. Here, then, is a further intuition. My intuition is that we are permitted to remove B_B from the atmosphere, but not to remove B_V. My suspicion is that this has little to do with the otherwise commonplace distinction between the natural versus the non-natural (B_V versus B_B), but more to do with the direction of agency.

Nevertheless, this is our puzzle. Why is this the appropriate solution? If it is appropriate to remove some B_B from the atmosphere, why not remove B to our heart's content? Why must we be more restrained? Or, put differently, if we can steer atmospheric concentrations in one direction, oughtn't we be permitted to steer them in the direction of an ideal state, supposing that we can identify what that ideal state might be? I think it is impermissible to steer atmospheric concentrations of A, B, and C toward an ideal state, and I think it has little to do with the question of whether we can identify an ideal state.

Consider this conflating observation: Suppose it is discovered that A and C are both more reactive with another element, β, than with B. We know that this element β is harmless, and that the combination of AβC is equally harmless. Suppose we introduce this harmless element β as a replacement for B. We'll then have the inert compound AβC. This would have the effect of rendering ABC inert, effectively removing the impacts on other populations. Now consider the following case.

Smokestack

Suppose we could construct a giant smokestack to emit β over the town of Incenter. Would it be permissible to shoot β into the atmosphere in order to pre-empt the bonding of B with A and C?

I suspect many would find such a resolution to be impermissible, mostly because it operates on the presupposition that one can alter the atmosphere (an environment with wide distributional influence and impact) in order to satisfy the concerns of the relative minority, the human population of Incenter. Moreover, it is not in any respect a true remediation solution. Rather, it involves moving the universe not from one state to a previous state (from state Y back to state X, say), but from one undesirable state to another, presumably more desirable, state (from state Y to state Z). It does so by way of intermingling β emissions with the lives of others, human and non-human. What should be clear is that the two cases differ primarily in that one is more or less of natural etiology while the other is more or less non-natural.[2] Consider, by contrast, the following case.

Counteraction

Acme and Capme set up some device to combine β with A and C, respectively, on site, in the factory, before A or C is emitted into the atmosphere. They then plan to emit compounds of Aβ and Cβ into the atmosphere, knowing that they will eventually combine to form AβC. Provided that β is harmless, as well as the combination AβC, intuition suggests that such activities would be harmless.

I suspect that Counteraction is nowhere near as problematic as Smokestack, even though, again, β emissions are equivalent. The reason for this rests with the suggestion that emissions constitute a sort of "moral trespass," a term that I borrow unapologetically from Mark Sagoff (2004). Altering an emission before its release and, say, mitigating harms from that emission does weigh on the overall assessment of its permissibility (but does not authorize its release, even when harms are completely eradicated), until the emission has been released, at which time there is more to consider. This is neither a factor of time, nor a question about where the emission is in the pipeline, but rather a question about who is affected and what they could or would accept. Consider, instead, the following case.

Beta Jets

Suppose β is a naturally occurring molecule. It is prevalent and abundant near the town of Incenter, thanks to a nearby lava field from which jets of β shoot up out of the hot magma. Lucky for Incenter, β has always been present in elevated concentrations, latching on to atoms of A and C, thereby thwarting a potential catastrophe.

Here we have a case in which the natural state of affairs is beneficial to the residents of Incenter, perhaps without their knowledge. Such a case probably is not far from reality. Who knows what untold horrors could have befallen us had it not been for the preemptive remediation technologies of Mother Nature? Ecosystem services have provided innumerable benefits and a comfortable sanctuary from our inconsiderate tendency to degrade and sully the Earth. Are we permitted to continue emitting A and C into the atmosphere in the face of such knowledge? One might think we are, since the Beta Jets have been around in abundance. I'm not clear on that. As I've tried to argue before, and as I believe I am justified in arguing, even emitting harmless (or perhaps beneficial) compounds into the atmosphere amounts to moral trespass. In order to authorize such actions, we need to solicit input from all who would be touched (not necessarily harmed) by our action. To see this, consider the following case.

Beta spritzers

Suppose that the lava field is discovered to be a natural source of β emissions, but that it is emitting only half as much as would be needed to offset the confluence of B emissions in the atmosphere. It is only "spritzing" β into the air. We could easily allow the lava field to release more β by drilling holes in the lava.

Are we permitted to drill holes in the lava? I don't think we are. I don't think we are for the same reason that I don't think we're permitted to build a smokestack that spews β into the atmosphere. Doing so involves asserting our control over the situation, injecting our culpability into nature's air passages, enforcing our will on the lives, activities, projects, and interests of others. If something unforeseen were to eventuate from our action—suppose β exhibits grue- or bleen-like properties (Goodman 1955), that it is harmless before December 31, 2049 but deadly after January 1, 2050—we can be said to have done wrong.

Permissible Interventions

Return, then, to Acme, Beatme, and Capme. If Acme emits A knowing that it will intermix with B and C and does nothing to stop the mixing of B and C, it is reasonable to suggest that Acme will have done something quite wrong. Most agree to this. But I have been suggesting that even if Acme takes steps to ensure, if and when B and C intermix, that the impacts are minimized, it is not clear if it has yet done what is required of it. What is required is that it consider whether emitting A is justified, by which I mean something other than whether, on balance, benefits outweigh costs. Acme must understand the full breadth of its action in determining how to evaluate whether and what sort of remediation technology to employ. Moreover, I have argued that actions by any of the three companies must attend to concerns about the direction and dissipation of agency, about the intermingling of agency with the interests of others. Acme is responsible for emitting A_A, whatever its effects. Beatme is responsible for emitting B_B, whatever its effects. So too for Capme. Their responsibility lies with their action—particularly with the reasons for their action—and not solely with the effects of their action. As their emissions dissipate into the atmosphere, their responsibility becomes entangled with the lives, actions, and decisions of others, including the other companies around them. Their responsibility then dissipates as it intertwines with the interests of others, such that there is no longer a simple question of turning back the clock, but only a question about how far back the clock is permitted to be turned.

This question about what is a permissible intervention, I think, can be answered only through deliberative engagement with all affected parties. There are, however, a few signals about the direction in which we should seek an answer.

An accurate description of any RT must understand the technology not just in terms of whether a resultant state is better than the one that precedes it, but also about the direction of agency, the intermingling of interests, and the reasons that have motivated the initiating action. It is wrong, for instance, to assume, because some damage has been done somewhere, that therefore the entire site is open for transformation. This is particularly true of the initiating action was unjustified, but it holds for justified actions as well. What emerges from this observation is that justificatory burdens are lighter for actions that seek to reverse a harm done than for actions that seek to patch over a harm by introducing a new state of the world. Consider this scenario:

Creation of Harm Moves universe from state X $\rightarrow$ state Y.

In the creation of a harm, an actor moves the universe from state X to state Y. That much is clear. The question for this chapter is which direction we are permitted to move from there. My suspicion is that it is permissible to undo a harm but impermissible to patch over a harm:

Undoing of Harm RT Action Φ_1 moves state Y $\rightarrow$ state X.

Patching over Harm RT Action Φ_2 moves state Y $\rightarrow$ state Z.

Here's why. To determine whether the widget production is permissible, which is the first and more primary question for an actor, the entire configuration must be assessed and evaluated, including antecedent and consequent conditions, preliminary and postliminary justifications, as well as the validity claims of all affected parties. It must be looked at in the context of all three companies, as well as the residents of Incenter, and it must be understood from the perspective of A, B, and C emissions. This is the nature of almost all ethical evaluations: They are tied tightly to our actions, to our agency. As our actions slip from our control, they impact and intermingle more and more with others.

Given the balance of justificatory burdens, it is likely more justified to say that harms should only ever be undone, lest one risk further wrongdoing, adding the caveat that harms are not permitted to be unilaterally undone—taken all the way back to the original state X—as they are not always unwelcome harms. Depending on one's perspective, in fact, they are not always harms. In the case of many RTs, we must take extra precautions to ensure that we are only undoing what has been done, and not adding insult to injury. To see this, consider cases in which a *benefit* is introduced. In such cases, one is permitted to remove external benefits only if the benefits have not yet intermingled with the interests of those affected. One can spray nitrogen fertilizer in the air over another's farm, for instance, and remove the fertilizer before it hits the ground, thereby depriving a farmer of potential external benefits; but one cannot remove nitrogen from the farmer's soil without consulting with the farmer, even if one is responsible for having put the nitrogen there. The reason for this is the same in cases of harms as in cases of benefits: As the impact of one's actions intermingle with the interests of others, the justificatory burden grows greater.

One technology that seems to meet the standard of a weaker justificatory burden is the technology known as air capture or atmospheric scrubbing. As my colleague Roger Pielke points out, by way of advocating for the technology, "the IPCC, both in its 2005 report on capturing and sequestering carbon dioxide [IPCC 2005] and in its 2007 Fourth Assessment Report [IPCC 2007][,] mentioned air capture only in passing" (Pielke 2009). Air capture is a selective remediation technology. If employed cautiously, it makes it possible that a single actor can take away the harm-causing dimension of her wrongdoing.

Given the strong justificatory hurdles that must be leapt to ensure justification of movement in one or the other direction, it is much more difficult to justify a move from state Y to state Z than it is to justify a move from state Y to state X. It is much easier, that is, to justify undoing a harm backward than it is to justify patching over a harm forward. For practical purposes, RTs that move in the direction of undoing are those that should be pursued.

To arrive at this conclusion, I have presupposed that the permissibility or impermissibility of an emission hangs on its justification. One cannot indiscriminately make a mess of things for the sake of making a mess, or for reasons that haven't met and passed tests of wide deliberative scrutiny. It is very often the case—particularly with carbon emissions, but also with other sorts of negative externality emissions—that emitting actions were not themselves justified (or were justified according only across a very narrow justificatory horizon). On this view, remediation technologies may make permissible some actions that were otherwise impermissible. If the justificatory procedure is left open enough to allow affected parties input into the determination of whether to move forward with the RT, then it may be permissible.

Conclusion

In order to determine which courses of environmental remediation are permissible, we need to adopt a position of respect for the rest of the world. This means ensuring that our actions are justified—that they could meet with the wide reflective scrutiny of all affected parties, human and nonhuman, living and non-living. More than anything, we must ensure that the actions we seek to counteract are "our actions." When commission of an action falls most directly on the shoulders of an agent, then that agent can claim direct jurisdictional control over that action. The tired and problematic distinction between natural and non-natural is, from the above analysis, largely irrelevant, as the permissibility of remediation actions is contingent more upon the direction of agency than upon some particular status of the world. Nevertheless, such a distinction may be a helpful folk-psychological device for categorizing actions, at least in lay discourse, and therefore perhaps even for this conclusion. As with all cases, the antecedent conditions and the reasons we offer for our action may make all the difference regarding permissibility.

As far as some natural-ish events are concerned: Often we are not *obligated* to undo what has happened, though in many cases we are *permitted* to undo what has happened. As a rule of thumb, we ought not to try to remedy our actions by way of introducing new states of affairs that *appear* to be natural or that appeal to particular past or future states of the world. It is one thing to plant a new forest for the purpose of sequestering carbon, and yet quite another to replace a downed forest with the objective of restoring the world to its original forested state. In instances in which planting more forest in order to return the atmosphere to its original state may be justifiable, what authorizes a community to do so is (a) the forestry practices that have preceded the decision to plant that forest and (b) the harm done to the atmosphere by other practices. If a community decides that it would like to alter the state of the atmosphere by planting a forest, it is far from a foregone conclusion that it is permitted to do so; at least, it is not clear that it would be permissible in a hypothetical universe where there has been no forestry. On the other hand, if the community decides to plant a forest for the sake of having a forest, it *may* do so, but only on the condition that it considers the impact of its actions on all affected parties. This is no different than building a factory that emits β, except insofar as one solution offers up trees and the other offers up widgets. The justification of the planting of that forest will depend on the extent to which the project is acceptable to those affected.

So far as some non-natural-ish actions are concerned: We can undo what we have done, but we cannot undo the wrong associated with what we have done. If we have wronged another, we are obligated to undo what harms we have done and to seek reconciliation with the other for the wrong. In some cases this is impossible, just as in some cases it is impossible to gain credit for a good deed done. If one party has been wronged by others, whether outsiders can or should intervene is an open question. This open question can be answered only by appeal to all affected. In other words, once an act is committed by an actor, it is permissible to return the world back from state Y to state X, or anywhere along a continuum between Y and X—where X is the base-line state and Y is the resultant state—given that once an actor has acted upon the world he has injected the world with agential control and culpability. But an actor cannot then legitimately move the world from state Y to state Z—where Z is a new third state—without also injecting further control into the world. Movement from state Y to Z is permissible only in consultation with those affected, just as planting a forest or building a factory is permissible only in consultation with those affected. Movement from state Y back to X is permissible only insofar as others will not be further affected by that movement.

Early in the chapter I mentioned that not all remediation technologies are created equal. At that point, I mentioned several proposed RTs. Employing what we have concluded, we must ask ourselves what are appropriate technologies. On the reasoning that I have offered, any geoengineering project that purports to move the world from Y to Z, such as ocean fertilization (Hale and Dilling 2011) or stratospheric sulfur injections (Brovkin et al. 2009), and not from Y back to X, is impermissible. These projects are far too comprehensive to be considered permissible, except in the most dire circumstances. In extraordinary and dire circumstances, the permissibility of shifting the world to state Z may be left on the table, but such justification will be difficult, if not impossible, to establish. The justificatory hurdles are enormous. Should we need to clean up the world, as we do now, in the face of anthropogenic climate change, we should take special precaution to ensure that we do not accidentally shift to state Z. Technologies such as atmospheric scrubbing—a technology that removes carbon dioxide, and only carbon dioxide, from the atmosphere—if employed expressly for the purpose of shifting the world back from state Y to state X, may be the only permissible direction to pursue.

On the reasoning that I have offered, it is most transparently permissible to clean up an emission before its release, though it is not always required.

Some emissions are unproblematic, and may even be desired by some parties, but their permissibility will be determined by the extent to which the affected community could or would assent to their release. Oxygen and carbon, if released into a non-saturated environment, may be these sorts of pollutants. In cases in which there is *post facto* knowledge of the devastating effects of our emissions, it is permissible to remove that which we have put into the atmosphere, but it does not remove the wrongdoing.

I have argued, effectively, that we are permitted to remove some pollutants that we have ourselves directly contributed, or to remove pollutants on behalf of those that others have contributed, but we are not permitted to remediate pollutants with a technology that involves transforming or adding something new to the pollutant to reduce its harmful nature, unless it can be demonstrated that all affected parties could or would assent to the remediating action. For this chapter, I have ignored questions of political jurisdiction—questions about whether I am permitted or obligated to remediate emissions from Smith or Jones—and instead chosen to focus on related questions about justifiability.

Notes

1. One may question this as an objective, particularly given my earlier insistence that pollution involves a form of moral trespass that is not adequately characterized in the language of harms. Nevertheless, acknowledging that remediation technologies only ever remediate harms and cannot turn back the clock on wrongs, it is a short jump to understand that the determination of what is a tolerable level of pollution must be arrived at through some alternative calculus or mutually respectful methodology.

2. I would like to avoid a lengthy discussion of what is natural versus what is non-natural. Instead, I prefer to understand the distinction between the two cases in terms of whether there is some agential involvement.

References

Boonin, D. 2003. *A Defense of Abortion.* Cambridge University Press.

Brovkin, V., V. Petoukhov, M. Claussen, E. Bauer, D. Archer, and C. C. Jaeger. 2009. Geoengineering climate by stratospheric sulfur injections: Earth system vulnerability to technological failure. *Climatic Change* 92: 243–259.

Buessler, K. O., S. C. Doney, D. M. Karl, P. W. Boyd, et al. 2008. Ocean iron fertilization—Moving forward in a sea of uncertainty. *Science* 319: 162.

Caldeira, K., and M. E. Wickett. 2005. Ocean model predictions of chemistry changes from carbon dioxide emissions to the atmosphere and ocean. *Journal of Geophysical Research Oceans* 110: 1–12.

Chisholm, S. W., P. G. Falkowski, and J. J. Cullen. 2001. Dis-crediting ocean fertilization. *Science* 294: 309–310.

Dennett, D. C. 1995. Intuition pumps. In *The Third Culture: Beyond the Scientific Revolution*, ed. J. Brockman. Simon and Schuster.

Department of Energy. 2008. Ocean Sequestration: Key R&D Programs and Initiatives. http://www.fossil.energy.gov

Dilling, L. 2007. Toward carbon governance: Challenges across scales in the United States. *Global Environmental Politics* 7: 28–44.

Dilling, L., S. C. Doney, J. Edmonds, K. R. Gurney, R. Harriss, D. Schimel, et al. 2003. The role of carbon cycle observations and knowledge in carbon management. *Annual Review of Environment and Resources* 28: 521–558.

Enkvist, P.-A., T. Nauclér, and J. Rosander. 2007. A cost curve for greenhouse gas reduction. *McKinsey Quarterly* 1: 35–45.

Goodman, N. 1955. *Fact, Fiction, and Forecast.* Harvard University Press.

Hale, B. 2009. What's so moral about the moral hazard? *Public Affairs Quarterly* 23: 1–26.

Hale, B., and L. Dilling. 2011. Carbon sequestration, ocean fertilization, and the problem of permissible pollution. *Science, Technology & Human Values* 36 (2): 190–212.

Hale, B., and W. Grundy. 2009. Remediation and respect: Do remediation technologies alter our responsibilities? *Environmental Values* 18: 397–415.

Herzog, H. 2003. Assessing the Feasibility of Capturing CO_2 from the Air. Publication No. LFEE 2003-002 WP. Laboratory for Energy and the Environment, Massachusetts Institute of Technology.

IPCC (Intergovernmental Panel on Climate Change). 2005. *Special Report on Carbon Dioxide Capture and Storage.* Cambridge University Press.

IPCC. 2007. *Working Group Iii: Mitigation.* Cambridge University Press.

Jamieson, D. 1996. Ethics and intentional climate change. *Climatic Change* 33: 323–336.

Jones, N. 2009. Climate crunch: Sucking it up. *Nature* 458: 1094–1097.

Keith, D. W., M. Ha-Duong, and J. K. Stolaroff. 2006. Climate strategy with CO_2 capture from the air. *Climatic Change* 74: 17–45.

Kintisch, E. 2007. An ethics code for ocean carbon experiments. *ScienceNow* (http://news.sciencemag.org).

McMahan, J. 2000. Moral intuition. In *The Blackwell Guide to Ethical Theory*, ed. H. LaFollette. Blackwell.

Nelson, J. A. 2008. Economists, value judgments, and climate change: A view from feminist economics. *Ecological Economics* 65: 441–447.

Parson, E. A. 2006. Reflections on air capture: The political economy of active intervention in the global environment. *Climatic Change* 74: 5–15.

Pielke, R. A., Jr. 2009. An idealized assessment of the economics of air capture of carbon dioxide in mitigation policy. *Environmental Science & Policy* 12: 216–225.

Potter, C., P. Gross, S. Klooster, M. Fladeland, and V. Genovese. 2008. Storage of carbon in US forests predicted from satellite data, ecosystem modeling, and inventory summaries. *Climatic Change* 90: 269–282.

Powell, H. 2007. Fertilizing the ocean with iron. *Oceanus* (http://www.whoi.edu).

Rawls, J. 1951. Outline of a decision procedure for ethics. *Philosophical Review* 60: 177–197.

Rickman, J. E. 2002. Imagine No Restrictions on Fossil-Fuel Usage and No Global Warming! News release, Los Alamos National Laboratory, April 9.

Sagoff, M. 2004. *Price, Principle, and the Environment*. Cambridge University Press.

Schneider, S. H. 2006. Climate change: Do we know enough for policy action? *Science and Engineering Ethics* 12: 607–636.

Scott, K. 2005. Day after tomorrow: Ocean CO_2 sequestration and the future of climate change. *Georgetown International Environmental Law Review* 28: 57–108.

Sencerz, S. 1986. Moral intuitions and justification in ethics. *Philosophical Studies* 50: 77–95.

Singer, P. 2006. Ethics and climate change: A commentary on MacCracken, Toman and Gardiner. *Environmental Values* 15: 415–422.

Spreng, D., G. Marland, and A. Weinberg. 2007. CO_2 capture and storage: Another Faustian bargain? *Energy Policy* 35: 850–854.

St. Clair, R. 1999. Commercial Ocean Fertilization: A Wise Use of Scientific Discovery? MIT News Office (http://web.mit.edu).

Stolaroff, J. K. 2006. Capturing CO_2 from Ambient Air: A Feasibility Assessment. Doctoral dissertation, Carnegie Mellon University.

Suthersan, S. S., and F. C. Payne. 2004. *In Situ Remediation Engineering*. CRC Press.

Thom, R. M., et al. 2002. Investigations into wetland carbon sequestration as remediation for global warming. In *International Conference on Wetlands and Remediation; Wetlands and Remediation II*, ed. K. Nehring and S. Brauning. Battelle Press.

Warner, S. D. 2007. Climate change, sustainability, and ground water remediation: The connection. *Ground Water Monitoring and Remediation* 27: 50–52.

6 Early Geoengineering Governance: The Oxford Principles

Clare Heyward, Steve Rayner, and Julian Savulescu

Geoengineering, defined by the United Kingdom's Royal Society as "the deliberate large-scale manipulation of the planetary environment to counteract anthropogenic climate change" (Royal Society 2009, 1), is attracting increasing interest. Until an article by Paul Crutzen (2006) propelled geoengineering into the scientific mainstream, it was regarded as disreputable and dangerous even to talk about (Lawrence 2006). Its place in the scientific establishment is virtually confirmed by the fact that the Intergovernmental Panel on Climate Change (IPCC) has included a review of geoengineering in its Fifth Assessment Report. The main reason for this increasing interest is concern over the lack of progress in reducing global CO_2 emissions, which are causing anthropogenic climate change. Geoengineering has been presented as a third option, alongside conventional mitigation and adaptation, to address global climate change.

It is widely acknowledged that geoengineering raises complex ethical, social, legal, and political questions and that in some cases these might be more difficult to resolve than the scientific questions (Rayner 2010). Development of powerful technologies and geoengineering will be no exception. Moreover, unlike other new technologies, such as biotechnology, genetic modification, and nanotechnology, geoengineering research will be set against the backdrop of the controversies and complexities encountered in the politics of climate change.

To date, there have been at least four sets of proposed principles for the governance of geoengineering. The Oxford Principles (Rayner et al. 2009) are one such set and arguably the most influential. In this chapter we outline the Oxford Principles and the societal values they were intended to capture and compare them with alternatives. In doing so, we elaborate on their intended function. First, it will be useful to elaborate on what geoengineering is and why it causes controversy.

Table 6.1

Comparison of proposed principles for the governance of geoengineering.

	Oxford Principles	Asilomar Principles	Morrow et al's principles	Jamieson's principles
Content	1. Geoengineering to be regulated as a public good 2. Public participation in decision making about geoengineering 3. Disclosure of geoengineering research and open publication of results 4. Independent assessment of impacts 5. Governance before deployment	1. Promoting collective benefit 2. Establishing responsibility and liability 3. Open and collaborative research 4. Iterative evaluation and assessment 5. Public involvement and consent	1. The Principle of Respect 2. The Principle of Beneficence and Justice 3. The Minimization Principle	1. Technical feasibility 2. Reliable prediction of consequences 3. Socio-economically preferable outcomes 4. No systematic violation of well-founded ethical principles, including the importance of democratic decision making, avoidance of irreversible change and the importance of learning to live with nature
Scope of applicability	Early research to any deployment	Field trial research	Field trial research	Deployment only
Type of principles	Procedural	Procedural	Procedural	Substantive ethical principles
Remarks		Requires clarification about level of public involvement: is it limited to consultation?	Beneficence and justice are different moral concepts; redundancy of Minimization Principle	Substantive ethical principles are included; therefore this list of principles does not represent the minimum consensus possible

Geoengineering: A Brief Overview

Anyone interested in "the governance of geoengineering" faces an unusual challenge in that there are two senses in which geoengineering does not yet exist. First, mature geoengineering technologies do not yet exist. There is no geoengineering technology that is ready for deployment, nor is there anything resembling a viable technological system, understood as comprising not only equipment but also maintenance, management, financing, and regulation. All that exists at present are components that might become part of a geoengineering technology. This poses an obvious challenge for any system of governance: It is not possible to construct detailed regulations for an "imaginary" technology; we cannot know how it will look, how it will function, or what benefits and problems it will create. Second, to talk of "the pros and cons of geoengineering" can misleadingly imply that all geoengineering technologies raise the same problems. This is not the case. Geoengineering is an umbrella term for an extraordinarily diverse range of technologies. The Royal Society listed a total of twenty three (Royal Society 2009). The current convention is to divide geoengineering into two categories: CO_2-removal technologies (CDR) and solar radiation management technologies (SRM), which increase albedo to reflect the sun's rays. The techniques listed range from the seemingly mundane (e.g., planting trees) to those that would not be out of place in a science-fiction novel (e.g., space mirrors). Some can be territorially contained, (e.g., enhanced weathering on land, bioenergy, sequestration); others can be effective only if released into global systems (e.g., ocean fertilization and stratospheric sulfate aerosol particle injection). Such a diverse set of technologies will all have their particular pros and cons and raise a wide range of governance challenges.

The variety of technologies currently being considered means that they engender a wide range of political and governance challenges (Rickels et al. 2011). We shall present a sample. Some have argued that geoengineering is hubristic. This hubris objection can take two forms. The first is that it is not humanity's place to manipulate the global climate in such a drastic way—humanity "does not have the right" to do so (e.g., Robock 2008). The second form of the objection points to the overbearing self-confidence that must be possessed by those who think they are capable of manipulating the climate (e.g., Gardiner 2011). Another ethical objection, complementary to the challenge of hubris, is that climate change is itself a wake-up call, that the human attitude toward the natural world is seriously wrong. On this line of argument, geoengineering is a continuation of this wrong-headed way of living. These are ethical objections—objections based on ideas of the good life.

Other objections are focused on issues of justice and legitimacy. One concern is that conducting research into climate geoengineering might pose what is known in the insurance industry as "moral hazard" (Baker 1996) by encouraging a relaxed attitude toward emissions reductions. Researchers and some politicians who have previously disputed the need for drastic mitigation (whether or not they accept that global warming is anthropogenic) are more enthusiastic about the prospect of some forms of geoengineering—particularly sulfate aerosol particle injection, which is regarded by some as relatively cheap (Barrett 2008; Weitzman 2009; Bickel and Lane 2009). If these voices are politically influential, support for mitigation might be undermined and this could result in future generations facing increased risk of adverse climatic changes should the technological gamble be unsuccessful. Questions of justice in siting of technical facilities will arise. Few people wish to live close to large industrial infrastructures. Furthermore, those who are made worse off from effects of testing and deployment (including unintended and unforeseen effects) might claim for fair compensation. These are questions of justice.

Other concerns relate to the legitimacy of the development and deployment of geoengineering and to social control over it. There is a possibility of "lock-in," which might take various forms. For example, some worry that once geoengineering research gathers momentum, it will continue to advance and we will proceed down the slippery slope until deployment becomes inevitable. Jamieson puts it this way (1996, 333): "[W]e seem to have a cultural imperative that says that if something can be done, it should be done. For whatever reason, technologies in this society often seem to develop a life of their own that leads inexorably to their development and deployment. Opposing the development of a technology is seen as 'Luddite'—an attempt to turn back progress that is doomed to failure." Social lock-in might also arise from the creation of vested economic interests. If a technology requires heavy initial investment, then those who provide the funds will lobby for the technology to continue to be developed and used. This could be the case with some forms of direct air capture. Some technologies are also liable to technical lock-in. One problem for stratospheric sulfate aerosol injection in particular is the termination effect. Should the program be discontinued for any reason, the result would be a rapid increase in global temperature (appropriate to the prevailing concentrations of atmospheric greenhouse gases), which would be harder to manage than any increase that would have occurred without deployment.

Some have suggested that a single state, or even perhaps a "Greenfinger" (Victor 2008), could deploy a technology unilaterally. Depending on the stated intentions of the actors, this could be seen as an act of aggression or of international civil disobedience. Whether these predictions come to pass will depend on the complex processes of international relations (Horton 2011; Millard-Ball 2012). It is, however, at least possible that the successful development of a geoengineering technology could be a game changer in global politics—and not only climate politics. Geoengineering will therefore raise issues of procedural justice, both within and between states.

Geoengineering research could be of great benefit if it were to contribute to averting climate impacts, or it could cause great harms and exacerbate existing injustices. These concerns cannot be overlooked if geoengineering research and deployment is going to be just and socially legitimate. Therefore, the Oxford Principles were proposed as a draft framework of guiding principles for the collaborative development of a system of geoengineering governance.

The Oxford Principles

The Oxford Principles were intended to capture widely held societal values that should be respected at all stages in the development of all forms of geoengineering technologies. However, they are a work in progress, and refinement and restatement will be necessary (Rayner et al. 2013). Some initial suggestions for how the Principles might be taken forward are made later in the chapter. The only two major assumptions that are not revisable are that at least some research into geoengineering is permissible in principle and that all geoengineering research and development should be subject to some sort of governance regime.

The original text of the five Oxford Principles is reproduced below, with some brief comments about the social value it was intended to capture.

- **Principle 1: Geoengineering to be regulated as a public good**

> While the involvement of the private sector in the delivery of a geoengineering technique should not be prohibited, and may indeed be encouraged to ensure that deployment of a suitable technique can be effected in a timely and efficient manner, regulation of such techniques should be undertaken in the public interest by the appropriate bodies at the state and/or international levels.

In an article elaborating on the values of the Oxford Principles, Rayner et al. (2013) explained that "the principle that geoengineering should be regulated as a public good acknowledges that all of humanity has a common interest in the good of a stable climate and therefore the means by which this is achieved. It suggests that the global climate must be managed jointly, and for the benefit of all, with appropriate consideration given to future generations."

Principle 1 has caused some confusion. For example, some commentators believed erroneously that emphasis on the "public" precluded a role for the private sector, or the granting of patents (Rayner et al. 2013). There are legitimate concerns about the role of the private sector and intellectual property, but Principle 1 does not automatically rule out either. What matters is not whether the private sector is involved, but where ultimate control rests. The provision of conventional, non-rivalrous, and non-excludable public goods can be done with involvement of the private sector. An example is national defense. Commercial companies have technical expertise and are heavily involved in the making of new equipment and technologies, but only the government can determine if, how, by whom, and to what purpose those technologies are to be used. The commercial sector is involved in the process, but the government is in ultimate control.

- **Principle 2: Public participation in geoengineering decision making**

> Wherever possible, those conducting geoengineering research should be required to notify, consult, and ideally obtain the prior informed consent of those affected by the research activities. The identity of affected parties will be dependent on the specific technique which is being researched—for example, a technique which captures carbon dioxide from the air and geologically sequesters it within the territory of a single state will likely require consultation and agreement only at the national or local level, while a technique which involves changing the albedo of the planet by injecting aerosols into the stratosphere will likely require global agreement.

Principle 2's requirement of public participation suggests a concern for social legitimacy (Rayner et al. 2013). From the explanatory text, it becomes clear that Principle 2 is an appeal to the all-affected principle, by which all those affected by a decision should have a say in its making (Whelan 1983). This principle is popular, but several issues will have to be addressed before it can be made operational.

First, the all-affected principle depends on a prior account of important interests (Heyward 2008). To state that any person who is affected in any possible way should have a say in the decision-making process would make the process far too unwieldy in order to be effective. It is therefore necessary to state what kinds of interests must be affected in order that an agent be entitled to a say in the decision.

Second, there is the question of whether the all-affected principle requires explicit consent. Securing explicit consent from all makes the decision-making process much more complicated. Although some argue that decisions must be justifiable to all affected, securing explicit consent is not always necessary. The Oxford Principles do not prescribe specific understandings of concepts such as affectedness and consent. Nor do they prescribe measures to assess such understandings. Different forms of consent—for example, revealed consent, hypothetical consent, and explicit prior consent (Rayner 1984)—might be appropriate, depending on the stage of technological development and the interests at stake. Political and legal cultures will affect the mode and the extent of public participation around the world. However, these structures must, under principle 2, be recognizably legitimate.

- **Principle 3: Disclosure of geoengineering research and open publication of results**

> There should be complete disclosure of research plans and open publication of results in order to facilitate better understanding of the risks and to reassure the public as to the integrity of the process. It is essential that the results of all research, including negative results, be made publicly available.

The third principle embodies the value of transparency. Prompt and complete disclosure of research plans and open publication of results should facilitate better understanding of the risks and should allow the public to be assured as to the integrity of the process (Rayner et al. 2013). Transparency is valuable instrumentally, in that access to relevant information about a proposal is obviously necessary if parties are to be able to consent. It is also valuable in itself. Even if one does not have a direct say over any particular matter, to be informed of decisions is an acknowledgment of one's moral status. In the absence of transparency, an agent is effectively "kept in the dark" and subject to exploitation or benign but disrespectful paternalism.

Disclosure creates a risk that agents could use the information to develop technologies for nefarious ends. These issues arise in scientific research, particularly research into diseases and pathogens, but worries about disclosure dual use do not always result in censorship. Moreover, it is important not to overstate these concerns. Terrorist acts have the most impact if they are quick, directly attributable, and have immediate and obvious effects, whereas interventions in the climate must be conducted on a long time scale and involve complex relationships between climatic changes and human impacts. Therefore, geoengineering does not appear attractive to terrorists. Admittedly, this might change if techniques were to become targeted and capable of creating immediate effects. Therefore, as with much else, decisions about disclosure will have to be made case by case. However, the burden of proof should fall on those who advocate restrictions.

- **Principle 4: Independent assessment of impacts**

An assessment of the impacts of geoengineering research should be conducted by a body independent of those undertaking the research; where techniques are likely to have transboundary impact, such assessment should be carried out through the appropriate regional and/or international bodies. Assessments should address both the environmental and socioeconomic impacts of research, including mitigating the risks of lock-in to particular technologies or vested interests.

Principle 4 states that regular assessments of the impacts of geoengineering research should be conducted by an independent body. Such assessments might be conducted by research organizations and funders, by regional or national governments, or even by appropriate international bodies (Rayner et al. 2013). Implementation of Principle 4 will require careful consideration of how to ensure that risk-assessment bodies are fully independent and careful consideration of the criteria for distinguishing whether an activity poses unacceptable risks. Here, the debates about the precautionary principle and debates about risk management and governance in other areas might be instructive. The precautionary principle has been interpreted in very different ways. The strong form of the precautionary principle holds that even a slight doubt about the safety of an activity can be grounds to preclude it (Sunstein 2005). As Sunstein argues, this effectively paralyzes activity; thus, we can assume that it would preclude geoengineering research activities. A weaker form of the precautionary principle holds that "a lack of decisive evidence of harm should not be ground for refusing to regulate" (ibid., 18). This coheres with the Oxford Principles: Because of a lack of research, there is no evidence that geoengineering research or deployment *will* cause harm but, because of the possible magnitude and spread of the potential harm, some oversight is justifiable. Rather than seeking to eliminate all risks involved in the development of geoengineering, the goal of any risk assessment should instead be to establish whether the risks are *reasonable* (Savulescu 1998). Reasonableness of risk concerns its level, its magnitude, and its distribution. Ensuring that risks are reasonable includes taking steps to assess all the current evidence for the level of and the magnitude of risk and to consider whether further research (undertaken by means less likely to have environmental impact) could better estimate the level of risk.[1] It also includes taking steps to minimize the identified risks and to assess whether the potential benefit, in terms of knowledge or welfare, is worth the risk (Savulescu and Hope 2010). Such assessments have the potential to include risk-reduction requirements and should contribute to public engagement (Rayner et al. 2013).

- **Principle 5: Governance before deployment**

  Any decisions with respect to deployment should only be taken with robust governance structures already in place, using existing rules and institutions wherever possible.

The fifth principle does not advocate eventual deployment, but simply indicates that any decision about whether or not to deploy must be made in the context of a strong governance structure. Although this might be built on existing institutional and legal arrangements for the management of scientific research, some geoengineering techniques might require new explicit international agreements or reforms of global governance institutions (Rayner et al. 2013).

Principles for Geoengineering Governance: A Brief Comparison

As was acknowledged earlier in the chapter, the Oxford Principles are not the only set of principles that have been proposed to guide the development of geoengineering. In this section we compare the Oxford Principles with their three main alternatives: those proposed by Jamieson (1996), those proposed by Morrow et al. (2009), and those proposed by delegates to the 2010 Asilomar Conference (ASOC 2010).

Long before geoengineering began to be taken seriously, Dale Jamieson stipulated that the following conditions must be met for an intentional climate-change project to be permissible: "(1) the project is technically feasible; (2) its consequences can be predicted reliably; (3) it would produce states that are preferable in socio-economic terms to alternatives; and (4) implementing the project would not seriously and systematically violate important, well-founded ethical principles or considerations" (Jamieson 1996, 326). The well-founded ethical principles Jamieson had in mind include the importance of democratic decision making, avoidance of irreversible changes, and the importance of "learning to live with nature." Excepting Jamieson's first principle, which states a necessary condition of technical feasibility, these principles allude to some of the societal values expressed in the Oxford Principles. Jamieson's second principle of reliable prediction of consequences is a stronger version of the fourth Oxford Principle, which requires prediction and assessment of impacts. The third of

Jamieson's principles holds that geoengineering must result in socially and economically preferable outcomes. That is, it asks what the appropriate goal of geoengineering should be—as does the first Oxford Principle. Jamieson does not state his view on what an economically and socially preferable outcome is, just as the Oxford Principles do not specify what counts as the benefit of all.

Regarding Jamieson's fourth principle, one of the "widespread ethical norms" that must be respected is the importance of democratic decision making. This is a clear acknowledgment of the importance of public participation. The second of the Oxford Principles highlights the importance of public participation in decision making, although the Oxford Principles are not pre-committed to Western democratic processes as the only means of legitimizing decisions.

However, another of the widespread ethical norms included in Jamieson's fourth principle is a substantive ethical principle: the importance of living with nature. Jamieson argues for its inclusion on the grounds that even if deployment of geoengineering was successful it would reinforce the view that the "proper human relationship to nature is one of domination" (Jamieson 1996, 332)—a view which he believes is dangerous in itself. Thus Jamieson draws on a particular view of nature and the proper human relationship to it. Using "nature" as the basis of political prescriptions is a common but flawed strategy and has led to counterproductive entrenchment and ad hominem arguments in the climate-change debates (Rayner and Heyward 2013).

To avoid problems in embodying one substantive ethical worldview about nature and human relationship, therefore, the Oxford Principles are entirely procedural and not substantive. This is not to say that ethical worldviews have no place in geoengineering debates, but rather that they are to be expressed in the public discussion which the Principles make space for, rather than being assumed at the outset. The lack of more controversial ethical claims might be advantageous because it could help make a geoengineering governance regime maximally inclusive.

Morrow et al. (2009) proposed another set of principles, inspired by the Belmont Principles for medical research involving human subjects. Their three principles are a Principle of Respect, a Principle of Beneficence and Justice, and a Minimization Principle. In the explanatory text they stated that the Principle of Respect should be understood as a requirement to secure the global public's consent before beginning empirical research. The Oxford Principles highlight the need for public participation, although they are more nuanced in that they hold that global agreement need not be necessary in all cases. They also refer to different forms of consent.

The Principle of Beneficence and Justice requires experimenters to achieve a favorable risk-benefit ratio while safeguarding human rights. The Minimization Principle requires that experiments take place over the smallest geographical area possible and have as little effect as possible on ecosystems, climate and human welfare. The Principle of Beneficence and Justice seemingly expresses a concern about the global public good, which will require thinking about principles of justice. Indeed, it goes beyond the Oxford Principles in suggesting that human rights must be safeguarded. The Minimization Principle is consonant with some of the interpretations of "reasonable risk" that could be part of the elaboration of the fourth Oxford Principle. Morrow et al. also propose that large-scale field experiments should not proceed without appropriate regulation. This could presumably be applied to deployment.

However, Morrow et al.'s proposal is less comprehensive than the Oxford Principles, and there are some puzzling aspects to their proposals. For example, it is questionable whether "the Principle of Respect" can be cashed out entirely in terms of consent. Additionally, beneficence and justice are different concepts, so it is unclear why they are combined into one principle. Moreover, there could be some duplication between this principle and the Minimization Principle. The Minimization Principle is basically an exhortation to reduce risk of harm as far as possible, and it seems that this would be a necessary condition of achieving a favorable risk-benefit ratio.

The Asilomar Recommendations (ASOC 2010, 9) are as follows:

1. promoting collective benefit
2. establishing responsibility and liability
3. open and collaborative research
4. iterative evaluation and assessment
5. public involvement and consent.

These recommendations explicitly draw on the Oxford Principles (ASOC 2010, 18), so it is not surprising that they are very similar in content. Recommendations 1, 3, and 4 highlight the same values as the first, third, and fourth of the Oxford Principles. However, there are some differences. Most notably, the Asilomar Recommendations do not have an equivalent of the fifth Oxford Principle, governance before deployment. Instead, the Asilomar Recommendations highlight the need to establish responsibility and liability for harmful effects of large-scale research activities related to climate engineering. If research proceeds to field testing, then this will certainly be a major concern, but this principle is less relevant to earlier stages of the research process. One other point to note about the Asilomar Recommendations is that the term "consent" appeared only in the name of the recommendation. In the explanatory text, "public involvement," "outreach," and "consultation" were used instead. Assuming that the Asilomar Recommendations are a work in progress, as the Oxford Principles are, we should not conclude that those who agreed to the Asilomar Recommendations take an instrumentalist view of public engagement (Stirling 2008). However, until this matter is clarified, the Oxford Principles have an advantage in highlighting consent rather than the narrower concept of consultation.[2]

From the above, it is clear that there are many similarities between the values expounded by the Oxford Principles and alternative principles for the governance of geoengineering. Themes of regulation for the common good, of public involvement and consent, and of careful attendance to the impacts of research projects are present in all.

One significant difference between the Oxford Principles and these three alternatives cannot be identified by looking at their content. This difference is the intended scope of application. The Oxford Principles have a much greater scope than the other three proposals. Jamieson's principles are seemingly intended to apply only to deployment. Morrow et al.'s proposal and the Asilomar recommendations are intended to cover only field tests conducted in the course of research. Morrow et al. deny that modeling carries any risks, which implies that it does not need governance. The Asilomar Conference report explicitly claims that existing scientific norms and standards are sufficient for modeling and for laboratory experiments (ASOC 2010, 25).

By contrast, the Oxford Principles are intended to apply to *all* stages of the research process. Research scientists are understandably reluctant to have to add to their administrative tasks, but governance at the earliest stage is still desirable. The reason for this is as follows: Even at the earliest stages, each research project will have a potential impact on future climate policy, whether that contributes to "normalizing" or "legitimizing" discourse on geoengineering, affects public perceptions of the seriousness of climate change, or sets the pathway for future research. Modeling and laboratory studies should be subject to a different form of regulation than field tests, but this is quite different from saying that early stages of research are exempt from a geoengineering governance system. In particular, given the dangers of hubristic claims being made about new technologies, and how some preliminary claims about the costs of sulfate particle injection have been interpreted outside of the scientific arena, the publication of all results, including negative results, is of just as much importance in the earliest stages of research as it is in the later stages.[3] Principles 2 and 4 are also relevant, although perhaps the concrete interpretation of them that would be appropriate to modeling and lab-based research does not require much change from standard research practice.

From Abstract Principles to Research Protocols

The Oxford Principles are abstract, high-level principles. They do not make concrete recommendations; they must be interpreted to fit a particular case. The authors intended them to be realizable in many different contexts and to be appropriate to the technology under consideration, its stage of development, and the wider social context of the research. The Oxford Principles set out the ways in which agents involved in geoengineering can be called to account. The precise action-guiding recommendations will be built up over time, through the use of research protocols for each stage of technological development. Before any activity, researchers should be required to prepare a research protocol explicitly articulating how the issues embodied in each of the Oxford Principles are to be addressed. This protocol will then be reviewed by a competent third party. The identity of the reviewing parties will be appropriate to the stage of research, but all must be invested with the authority to withhold approval until they are sure that the experimental design for that stage satisfies the Oxford Principles. Through the development of these protocols, the Principles will be translated into specific content, recommendations, and regulations appropriate to different technologies.

To start this process of elaboration, we suggest some ways in which the first four of the Oxford Principles might be translated into action-guiding recommendations. With respect to Principle 5, at this stage, it is hard to conceive of what "governance before deployment" might involve, so we refrain from speculation.

The first action-guiding specification of Principle 1 might be as follows: All intellectual-property interests are declared at the beginning of a research project. This recommendation follows the cancellation of the Stratospheric Particle Injection for Climate Engineering (SPICE) test-bed project in 2012 after the discovery that two researchers involved in the project had applied for a patent on the technology being tested. Other partners of the project (including the principal investigator, Matt Watson) had not been aware of this; Watson canceled the test-bed experiment as a largely pre-emptive measure. Patent application is common in scientific research, and Principle 1 does not preclude it. However, in order to be able to regulate geoengineering as a public good and to maintain a climate of trust between researchers, and between researchers and the public, it is necessary to know what forms of intellectual property are held, or being sought, and by whom. It could eventually turn out that there is a hypothetical case that all potential conflict of interests, not only intellectual-property interests, should be disclosed. Additionally, it could well be appropriate to disclose the identity of a project's funders, or whether researchers have relationships with commercial companies. This is similar to the declaration of potential conflict of interest that researchers must make before submitting academic articles or grant proposals.

The time has passed when science can progress without involvement and support of the wider public. No scientific research takes place in a social vacuum, and that this is especially true of geoengineering. The case for geoengineering has been presented in terms of developing solutions to the problem of anthropogenic climate change, and therefore the case for progressing with research must be considered in this wider context. Advocates of geoengineering research point to its global social benefits, but in doing so they must respect the fact that others may have different but reasonable views on what ultimately benefits the world.

Should small-scale experiments in geoengineering come to take place in the field, anyone who stands to be affected should be informed and consulted, and public participation should take place a reasonable length of time before the design of the experiments is finalized. This gives members of the public time to consider the proposals and raise concerns not only allowing for the possibility that some concerns could be factored into the experimental design, but acknowledging that public participation is not just another bureaucratic requirement to be ticked off the list before the "real research" takes place. Nor must any public engagement be regarded as a public relations exercise. In order for there to be a genuine debate on the merits of any geoengineering technology, expert researchers should take care not to engage in "stealth advocacy" (Pielke 2007) and should examine both the information provided to enable public participation and the consultative structures used to ensure that the debate is not closed down, and that alternative views can be put forward and investigated (Stirling 2008).

In addition, it has been suggested that there should be a registry, or an international database (Blackstock 2012, 159; UK Government 2010, 6), of geoengineering research projects. This is similar to the proposal that all clinical trials should be registered and their results, even if negative or disadvantageous to drug company sponsors, made publicly available (Savulescu et al. 1996). A registry prevents publication bias and harm to participants in clinical research. In the case of geoengineering, we are all potential participants in trials. This would be one way of beginning to implement Principle 3 (and would help implement Principle 2—providing information to the public is a necessary condition of public participation). The registry should announce planned experiments and trials to help ensure that any "inconvenient" results do not simply disappear (McGoey and Jackson 2009, 112) and thus prevent publication bias.

Finally, following from the Lohafex experiment conducted in the Southern Atlantic in 2009, Principle 4 might lead to the formulation of a requirement to provide risk assessments before any research project as well as assessment of impacts afterwards. Initially, the research team did not provide a prior risk assessment, believing that it was unnecessary and not legally required (Kintisch 2010). The German government then required the team to provide an assessment, delaying the experiment for some weeks. There should also be a complementary requirement that the statements presented in risk assessments are clearly consistent with those relating to the initial justification of the research project. For example, if the justification of doing research is the need for large-scale experiments in oceanic waters, then emphasis on the small scale of the experiment in the risk assessment and on the "coastal nature of the waters" will invite questions (e.g., Kintisch 2010, 159).

These are suggestions, nothing more, made simply to kick-start the debate. We invite criticism of them and encourage refinement and development of the Oxford Principles and the general development of technology-specific research protocols. Many have acknowledged that geoengineering raises ethical, social, and political issues. It is now time to start in-depth discussion of those issues.

Notes

1. For example, some critics of the SPICE test bed thought that more modeling should be done to see whether sulfate-particle injection was worth pursuing before field tests of delivery mechanisms were conducted.

2. To be clear, as the comments about Principle 2 state, consent might take many forms and might not include active and explicit consent. However, it is important that this understanding of consent is not ruled out in the text of the Oxford Principles. The concern is that by talking of public consultation, outreach and involvement, the explanatory text of the Asilomar Principles implies that the authors take

the view that the public should be able to *influence*, but not ultimately to *decide* on whether a research project goes ahead.

3. The low cost estimates of sulfate-particle injection have led to its being recommended by some economists (see Levitt and Dunbar 2009) and to a widely cited endorsement from Newt Gingrich (Vidal 2011).

References

ASOC (Asilomar Scientific Organizing Committee). 2010. The Asilomar Conference recommendations on principles for research into climate engineering techniques. www.climate.org/PDF/AsilomarConferenceReport.pdf

Baker, Ted. 1996. On the genealogy of moral hazard. *Texas Law Review* 72: 237–292.

Barrett, Scott. 2008. The incredible economics of geoengineering. *Environmental and Resource Economics* 39: 45–54.

Bickel, Eric, and Lee Lane. 2009. The Copenhagen Consensus: An Analysis of Climate Engineering as a Response to Climate Change. Copenhagen Consensus Center.

Blackstock, Jason. 2012. Researchers can't regulate geoengineering alone. *Nature* 486: 159.

Crutzen, Paul. 2006. Albedo enhancement by stratospheric sulfur injections: A contribution to resolve a policy dilemma? *Climatic Change* 77: 211–220.

Gardiner, Stephen M. 2011. *A Perfect Moral Storm*. Oxford University Press.

Heyward, Clare. 2008. Can the all-affected principle include future persons? Green deliberative democracy and the non-identity problem. *Environmental Politics* 17: 625–643.

Horton, Josh. 2011. Geoengineering and the myth of unilateralism: Pressures and prospects for international cooperation. *Stanford Journal of Law, Science and Policy* 4: 56–69.

Jamieson, Dale. 1996. Ethics and intentional climate change. *Climatic Change* 33: 323–336.

Kintisch, Eli. 2010. *Hack the Planet: Science's Best Hope—or Worst Nightmare—for Averting Climate Catastrophe*. Wiley.

Lawrence, Mark. 2006. The geoengineering dilemma: To speak or not to speak. *Climatic Change* 77:245–248.

Levitt, Steven, and Stephen Dunbar. 2009. *Superfreakonomics: Global Cooling, Patriotic Prostitutes, and Why Suicide Bombers Should Buy Life Insurance*. Allen Lane.

McGoey, Linsey, and Elizabeth Jackson. 2009. Seroxat and the suppression of clinical trial data: Regulatory failure and the uses of legal ambiguity. *Journal of Medical Ethics* 35: 107–112.

Millard-Ball, Adam. 2012. The Tuvalu Syndrome. *Climatic Change* 110: 1047–1066.

Morrow, David R., Robert E. Kopp, and Michael J. Oppenheimer. 2009. Toward ethical norms and institutions for climate engineering research. *Environmental Research Letters* 4: 1–8. doi: 10.1088/1748-9326/4/4/045106

Pielke, Roger A. 2007. *The Honest Broker: Making Sense of Science in Politics*. Cambridge University Press.

Rayner, Steve. 1984. Disagreeing about risk: The institutional cultures of risk management and planning for future generations. In *Risk Analysis, Institutions and Public Policy*, ed. Susan Haddon. Associated Faculty Press.

Rayner, Steve. 2010. The geoengineering paradox. *Geoengineering Quarterly* (http://www.greenpeace.to/publications/The_Geoengineering_Quarterly-First_Edition-20_March_2010.pdf).

Rayner, Steve, and Clare Heyward. 2013. The inevitability of nature as a rhetorical resource. In *Anthropology and Nature*, ed. Kirsten Hastrup. Routledge.

Rayner, Steve, Clare Heyward, Tim Kruger, Catherine Redgwell, Nick Pidgeon, and Julian Savulescu. 2013. The Oxford Principles. *Climatic Change* 121: 499–512.

Rayner, Steve, Catherine Redgwell, Julian Savulescu, Nick Pidgeon, and Tim Kruger. 2009. Memorandum on draft principles for the conduct of geoengineering research. http://www.geoengineering.ox.ac.uk/oxford-principles/history/

Rickels, Wilfrid, Gernot Klepper, Jonas Dovern, Gregor Betz, Nadine Brachatzek, Sebastian Cacean, Kirsten Gussow, et al. 2011. Large-scale Intentional Interventions into the Climate System? Assessing the Climate Engineering Debate. Scoping report conducted on behalf of the German Federal Ministry of Education and Research (BMBF). Kiel Earth Institute.

Robock, Alan. 2008. 20 reasons why geoengineering may be a bad idea. *Bulletin of the Atomic Scientists* 64: 14–18.

Royal Society. 2009. Geoengineering the Climate: Science, Governance and Uncertainty. Policy document 10/09.

Savulescu, Julian. 1998. Safety of participants of non-therapeutic research must be ensured. *British Medical Journal* 16: 891–892.

Savulescu, Julian, and Tony Hope. 2010. Ethics of research. In *The Routledge Companion to Ethics*, ed. John Skorupski. Routledge.

Savulescu, Julian, Iain Chalmers, and Jennifer Blunt. 1996. Are research ethics committees behaving unethically? Some suggestions for improving performance and accountability. *British Medical Journal* 313: 1390–1393.

Stirling, Andy. 2008. "Opening up" and "closing down": Power, participation, and pluralism in the social appraisal of technology. *Science, Technology & Human Values* 33: 262–294.

Sunstein, Cass. 2005. *Laws of Fear: Beyond the Precautionary Principle*. Cambridge University Press.

UK Government. 2010. Government Response to the House of Commons Science and Technology Committee 5th Report of Session 2009–10: The Regulation of Geoengineering.

Victor, David. 2008. On the regulation of geoengineering. *Oxford Review of Economic Policy* 24: 322–336.

Vidal, John. 2011. Geoengineering: Green versus greed in the race to cool the planet. *The Observer*, July 9, 2011 (http://www.guardian.co.uk/environment/2011/jul/10/geo-engineering-weather-manipulation).

Weitzman, Martin. 2009. On modelling and interpreting the economics of catastrophic climate change. *Review of Economics and Statistics* 91: 1–19.

Whelan, Frederick. G. 1983. Democratic theory and the boundary problem. In *Liberal Democracy*, ed. J. Roland Pennock and John W. Chapman. New York University Press.

7 Design for Sustainability

Ibo van de Poel

Technology affects the environment in a number of ways. Some technologies contribute to the pollution of the environment; others lead to the use and possible exhaustion of nonrenewable natural resources, such as coal and uranium. Technology also contributes to environmental problems such as the greenhouse effect, overfishing, and loss of ecosystems. In the meantime, technology can contribute to the solution or prevention of many environmental problems. Innovative technologies may result in a reduction of energy consumption and in cleaning up environmental pollution, and capturing and storing CO_2 may contribute to avoiding or at least reducing the greenhouse effect.

Whether a particular technology increases or solves environmental problems depends in part on its type. Coal plants, for example, have a larger environmental impact than pencils. However, the environmental impact of a given kind of technology also depends on its design. Some refrigerators, for example, consume less energy than others. Copying machines that have double-sided printing as default option, rather than single-sided printing, are likely to reduce the consumption of paper. How a particular technology is designed thus matters for its environmental impact. For that reason, it makes sense to design for sustainability.

That engineers have a responsibility for the environment is now increasingly recognized. The code of conduct of the US National Society of Professional Engineers, for example, states that "engineers are encouraged to adhere to the principles of sustainable development in order to protect the environment for future generations" (NSPE 2010). The Code of Conduct of the European Association of National Engineering Societies (FEANI) states that "engineers ... shall carry out their tasks so as to prevent ... avoidable adverse impact on the environment" (FEANI 2012). These broadly formulated responsibilities also imply that engineers have a responsibility to design technological products for sustainability.

In recent years, design for sustainability has become increasingly popular and a number of handbooks have appeared (Stitt 1999; Birkeland 2002; Bhamra and Lofthouse 2007). Most of these handbooks offer practical design guidelines and tools such as life-cycle analysis (LCA) without discussing the notion of sustainability and its normative and contestable character in much detail. When I say that sustainability is a contestable notion, I do not mean that the desirability of sustainability as such is contested, but rather that exactly what we mean by "sustainability" is contested. This is witnessed by the number of definitions of "sustainability" that are now around.

The aim of my chapter is to do justice to the normative and contestable character of sustainability while also making the notion relevant in a practical sense for engineering design. I begin by arguing that we should see sustainability as a compounded value that consists of a range of other values, including intergenerational and intragenerational justice and care for nature. I then argue that, although sustainability is essentially a contestable notion, that does not make it impossible to design for sustainability. I then turn to the question of how we can translate the general value of sustainability into more specific design requirements that can guide the design of new technologies. I illustrate this with the case of biofuels, a case that I will also use to discuss different ways to deal with conflicts among the design requirements derived from the striving for sustainability. I then return to the contestable nature of sustainability and how that affects dealing with value conflicts in design for sustainability.

Sustainability as a Compounded Value

I conceive of sustainability not just as a technical notion but as a *value*. Values do not express preferences or things we want to attain just for ourselves, but things that we think are worth striving for in general. Values are normative, in the sense that they express what is good and desirable to attain. To say that sustainability is a normative notion is to say that it refers to the goodness of a certain matter. In the case of sustainability, the matter to which reference is made is usually a process of development. This is why discussions about sustainability are often cast in terms of sustainable development.

What is sustainable development? When is a certain development to be called sustainable? The most cited and probably the most influential definition of sustainable development has been provided by the World

Commission on Environment and Development (also known as the Brundtland Commission):

> Sustainable development is development that meets the needs of the present without compromising the ability of future generations to meet their own needs. It contains within it two key concepts:
>
> • the concept of needs, in particular the essential needs of the world's poor, to which over-riding priority should be given; and
> • the idea of limitations imposed by the state of technology and social organization on the environment's ability to meet present and future needs. (WCED 1987)

It can be argued that this definition of sustainable development refers to two types of justice: intergenerational justice, "without compromising the ability of future generations to meet their own needs," and intragenerational justice, "the essential needs of the world's poor, to which over-riding priority should be given."

The term "justice" is sometimes used in a broad sense to refer to what is morally good or desirable. In this usage, "just" is more or less equivalent to "morally right." My usage of "justice" in this chapter will be more specific. Following Barry (1999) and others, I will reserve the term "justice" for cases in which *distributive* considerations are at play. Intergenerational justice, so understood, is about the distribution of certain goods between generations, and intragenerational justice is about the distribution of goods within a generation (for example, between different parts of the population, or between countries). What goods are to be distributed and exactly what a just distribution of these goods entails are subsequent questions.

What, exactly, is the relationship between sustainability and intergenerational justice? If "intergenerational justice" refers to the just distribution of some X among generations, "sustainability" could be understood as referring to the *sustenance* of some X for the future. Whatever exactly X is, and whatever we exactly mean by "a just distribution," it appears that sustainability ("the sustenance of X") is at least a minimal condition for intergenerational justice ("the just distribution of X over generations"); if no X is sustained, X cannot be fairly distributed among generations.

Two caveats are in order with respect to this relationship between sustainability and intergenerational justice. First, a just distribution should exclude the possibility that no X is left for the next generation. In this sense, the relationship is not completely independent from what we mean by a just distribution. Second, it would seem that sustainability is a necessary but not a sufficient condition for intergenerational justice. If we sustain X for the next generation, there is no guarantee that X is justly distributed. It can, for example, be unjustly distributed between different parts of the population or between men and women. The latter are, however, cases of *intragenerational* injustice, whereas we are contemplating intergenerational justice here. It is not obvious that intergenerational justice would require more than sustaining the right amount of X. Of course, what amount of X to sustain for the next generation depends on what we mean by "a just distribution among generations." But once an amount has been determined, no more than sustaining it for the next generation seems required. It seems that, in practice, sustainability would be enough to bring about intergenerational justice.

Sustainability is not only a necessary but also a sufficient condition for intergenerational justice. There are at least three ways to conceive of the relationship between sustainability and intergenerational justice. One way is to see intergenerational justice as more fundamental than sustainability. It provides the reasons why we strive for sustainability, and it justifies sustainability. Why do we want to attain sustainability? Because it brings about intergenerational justice. A second possibility is to conceive of sustainability and intergenerational justice as more or less synonymous. A third option is to see intergenerational justice as one of the values constituting sustainability, but not necessarily as the only one.

I choose the third option here. The main reason for this choice is that more is at stake in policy and societal debates about sustainability than intergenerational justice. One other value is *intragenerational* justice as testified in the Brundtland Commission's definition. As we saw above, intergenerational justice can be achieved without intragenerational justice being achieved. Below, I will discuss an example, biofuels, that makes the conflict between intergenerational and intragenerational justice even clearer. Following the Brundtland Commission, among others, I will take it that developments that attain intergenerational justice but ignore considerations of intragenerational justice are not sustainable.

Another issue with respect to sustainability is whether it should be understood in anthropocentric or in biocentric terms. If it is understood in anthropocentric terms, sustaining nature and the environment are sought for the sake of human well-being; on a bio-centric view, nature is attributed final value. It is questionable whether this alleged final value of nature can be captured by the value of intergenerational justice. Like Barry, I am inclined to believe that "justice and injustice can be predicated only of relations among creatures who are regarded as moral equals in the sense that they weigh equally in the moral scales" (Barry 1999, 95). This implies that, although justice applies to relations and distributions among human beings, it does not apply to relations or distributions between human beings and other creatures or nature. This is not to deny that human beings can have moral obligations to animals or plants. It is merely to postulate that such moral obligations cannot be captured by the value of justice. Insofar as such moral obligations are thus considered part of sustainability, they cannot be captured by the values of intergenerational and intragenerational justice. I therefore take it that what I will call "care for nature" is the third value—in addition to intergenerational and intragenerational justice—constituting sustainability.

It should be noted that if X is understood in anthropocentric terms, it is not obvious that sustainability requires the protection of the environment or of nature. That is the case only if the sustenance of human opportunities or human welfare requires the protection of the environment or nature. If one assumes that environmental resources are renewable or can be replaced by man-made resources, sustaining human opportunities or welfare does not necessarily imply the protection of the environment. Gas and oil may be replaced by other sources of energy that are renewable. Even if certain environmental resources are renewable or replaceable, hardly anyone seems to assume the complete substitutability of man-made resources for environmental resources (Dobson 1998, 42) Therefore, in practice, anthropocentric notions of sustainability also imply the protection of the environment.

If one takes a biocentric view, X would not be restricted to *human* needs, opportunities, or welfare. It would also include parts of nature that are considered valuable in themselves. Think of the sustenance of endangered species, valuable ecosystems, or natural landscapes such as the Grand Canyon. On such a view, X is not just a metric from which the value and importance of other resources that are needed to sustain X can be derived; rather, X itself consists of a range of issues that each are valuable for their own sake.

Such a broad and multifaceted view of X does not necessarily require a biocentric perspective. If we think of X in broader anthropocentric terms—not in terms of needs, opportunities, or economic welfare, but rather in terms of the broad notion of *human well-being*—there are good reasons to assume that X also should be understood as multifaceted. The reason for this is that human well-being is itself a composed value that consists of a range of other values (Griffin 1986; Norton 1999; Nussbaum 2000; van de Poel 2012). Human well-being does not require only the fulfillment of physical needs. It also requires, for example, freedom, the capability to direct one's life individually and collectively and the ability to enjoy certain experiences, which in turn would probably require the sustenance of some parts of nature (e.g., certain species, ecosystems, or beautiful landscapes). On such a view, sustaining freedom or democratic institutions would be part of sustainability. It can even be argued that attaining freedom, democratic institutions, or intragenerational justice in cases where they do not exist yet is part of sustainability. This is also suggested by the Brundtland Report, which considers the abatement of world hunger to be part of sustainable development (WCED 1987).

We have seen that sustainability might be interpreted as the sustenance of some X over time and that this interpretation still leaves room for wide range of more specific conceptions of sustainability, ranging from ones that understood X as the fulfillment of human needs, or in terms of human opportunities, to multifaceted views that include in X not just a range of parts of nature but also important social institutions, among them democracy and freedom. Although intergenerational justice is an important motivating and justifying value for sustainability, sustainability is more than just the attainment of intergenerational justice. It also includes the values of intragenerational justice and care for nature.

Sustainability as a Contestable Concept

Even if there is considerable agreement that sustainability can be understood as the sustenance of some X, there is considerable disagreement about what the X to be sustained is and about what its values are. This disagreement often frustrates effective policy or, in our case, effective design for sustainability. The solution is usually sought in a definition of sustainability that attempts to resolve the disagreements. I would, however, like to suggest that this strategy is misguided.

It seems obvious that a definition alone cannot resolve substantial disagreements. The disagreements underlying the question of how to understand sustainability concern questions such as whether nature or the environment has final value and whether we should conceive of human well-being in terms of economic welfare or whether we should conceive of it more broadly to include freedom and democracy. The debate about sustainability is a debate about what kind of society we want to live in. This is a political debate than cannot be resolved by means of a definition.

Sustainability or sustainable development is what Michael Jacobs (1999) calls a "contestable concept," including it among other ethical-political concepts such as liberty, democracy, and justice. Such contestable concepts have, according to Jacobs, two levels of meaning. At the first level, there is agreement on the desirability of the concept, but it is defined abstractly or even vaguely—in the case of sustainability, perhaps as the idea that some X should be sustained for future generations. At the second level, there may be different *conceptions* of the general concept. In our case, there are different conceptions of sustainability, and they result in different answers to the question of what (that is, which X) is to be sustained, and also differences with respect to why and how this is to be sustained and different assumptions about the substitutability between human-made and natural capital (Dobson 1998).

As Jacobs writes (1999, 26), "At the second level there is contestation. This shouldn't be perceived a remediable lack of what sustainable development means: rather, such contestation *constitutes* the political struggle over the direction of social and economic development." Contestation at the second level is thus irresolvable, at least with respect to a *general* agreement on what we mean by "sustainability."

Given the different meanings of sustainability, one might wonder how design for sustainability is still possible. The answer to this question is threefold. First, in many cases design for sustainability will aim at taking away existing "unsustainabilities." People with different conceptions of sustainability might well agree about what is unsustainable about the current situation. More generally, strategies such as increasing energy efficiency and closing material cycles will be justified on a large range of conceptions of sustainability. Second, even if there is disagreement about exactly what "design for sustainability" means, the notion of sustainability still provides an overarching and common argumentative framework. This framework will not force agreement; however, it will help the contestants in the debate to judge arguments, and it might thus help them to come to agreement. Of course agreement is not guaranteed, but the shared concept of sustainability is helpful even if it does not directly resolve conflicts by a shared conception. Third, to design for sustainability requires a rather detailed understanding of what "sustainability" means only in the specific case at hand. This detailed understanding will always require a translation (or specification, as I will call it) of the general concept of sustainability to the case at hand. Agreement on a conception of sustainability does not guarantee agreement on the specification of sustainability in a specific case. On the other hand, disagreement about the conception of sustainability does not rule out agreement on how sustainability is to be understood in a specific case.

Sustainability is a contestable concept. Although there is agreement on the desirability of sustainability, there is substantial disagreement about the exact conception of it. This disagreement will not be resolved easily, because it represents fundamentally conflicting normative views on society and nature. However, this disagreement should not frustrate design for sustainability, as a shared conception of sustainability is neither a necessary nor sufficient condition for agreement on exactly what design for sustainability entails in a concrete case. What is needed for design for sustainability is, rather, a shared specification of it in a given case.

Specifying Sustainability

To design for sustainability we need more than an understanding of the value of sustainability and its compounding values, intergenerational justice, intragenerational justice, and care for nature. We need to specify sustainability in terms of a range of more specific design requirements. To help make such translations, I have proposed the concept of a *values hierarchy* (van de Poel 2013). A values hierarchy consists of three layers (figure 7.1). The upper layer consists of values, such as sustainability, intergenerational justice, intragenerational justice, and care for nature. The second layer consists of norms that prescribe, recommend, forbid, or discommend certain actions or options. The third layer consist of design requirements that describe desirable characteristics of the system to be designed.

A values hierarchy may be constructed from the bottom up as well as from the top down. From the top down, one begins by identifying the relevant values, which are then translated into more general norms and, eventually, into specific design requirements. This is what I have called

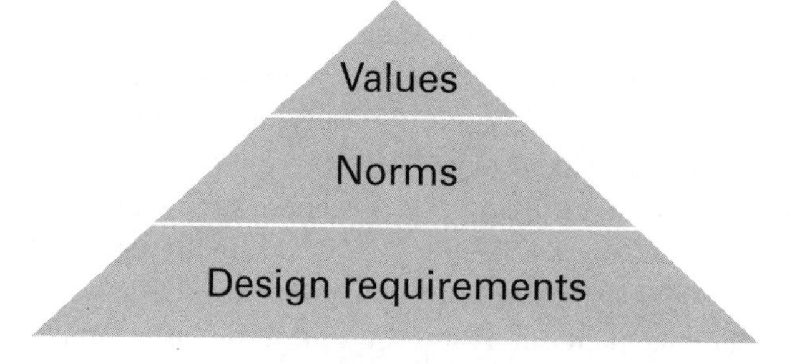

Figure 7.1
The three basic layers of a values hierarchy. Note that each of the layers may itself be hierarchically layered.

specification. Specification is context-dependent because it requires an interpretation of what a general value or norm means in a specific context. Usually more than one specification is tenable. This does not mean that any specification is possible or that all possible specifications are equally good. Nevertheless, the relevant values and norms can usually be specified in a number of defensible ways. The challenge is to judge if a specification meeting the lower-level design requirements would count as meeting the general values *in this specific context.*

If one constructs a value hierarchy from the bottom up, lower-level elements in a values hierarchy are done for the sake of higher-level elements. If a values hierarchy is constructed from the bottom up, the question to be asked is "For the sake of what is this design requirement aiming at?" The "for the sake of" relation is asymmetrical: If A is done for the sake of B, B is usually not done for the sake of A. We might, for example, formulate the goal of increasing fuel efficiency as a design requirement for a car for the sake of sustainability, but it would be nonsensical to say that we strive for sustainability for the sake of increasing the car's fuel efficiency.

The idea of a values hierarchy can be illustrated by the example of biofuels. Biofuels have become popular in recent years as a possible way to deal with an expected shortage of fossil fuels and as a way to reduce emissions of greenhouse gases. They have also been criticized because their environmental benefit are not always clear (Zah et al. 2007) and because they compete with food production, thereby driving food prices up and contributing to hunger and poverty, especially in the Third World (Naylor et al. 2007).

In 2011 the UK Nuffield Council on Bioethics proposed an ethical framework for the development of biofuels that includes the following five ethical principles:

> 1. Biofuels development should not be at the expense of people's essential rights (including access to sufficient food and water, health rights, work rights and land entitlements).
> 2. Biofuels should be environmentally sustainable.
> 3. Biofuels should contribute to a net reduction of total greenhouse gas and not exacerbate global climate change.
> 4. Biofuels should be developed in accordance with trade principles that are fair and recognize the rights of people to just reward (including labor rights and intellectual property rights).
> 5. Costs and benefits of biofuels should be distributed in an equitable way. (Nuffield Council on Bioethics 2011)

I will use these principles to construct a values hierarchy for biofuels (figure 7.2). I have chosen to adopt the principles for the context of the design and development of biofuels. Moreover, I have chosen to take sustainability here as the overarching value and intergenerational justice, care for nature and intragenerational justice as the values of which sustainability is compounded, as discussed above.[1] In figure 7.2 each of these compounding values is translated into one or more general norms that are relevant in the case of biofuels and that help to attain the relevant value.

In the case of intragenerational justice, one general norm is the sustenance of availability of fuels: Biofuels offer an alternative to fossil fuels which will be exhausted at some point in the future. In addition, they may help decrease the emission of greenhouse gases that is caused by fossil fuels. (See the third principle of the Nuffield Council.) Third, they should not increase other environmental problems. (See the second principle of the Nuffield Council.) The value of care for nature is translated into the maintenance of biodiversity, which the growing and harvesting of crops for biofuels potentially puts at risk as more land is used for agriculture production. The value of intragenerational justice relates to the first, fourth, and fifth principles of the Nuffield Council. In light of the first and fifth principles, it is undesirable that food prices increase as a result of the use of biofuels, as this particularly affect the world's poor. In addition, one might want to require that biofuels help to increase intragenerational justice by offering developing countries new development opportunities and new means of income. The fourth principle of the Nuffield Council is understood here as a general norm: Ensure just reward. Each of the general norms may, in turn, be translated in a number of more specific design requirement, as illustrated in figure 7.2.

Currently, there are no biofuels that meet all design requirements mentioned in figure 7.2 (Inderwildi and King 2009; Brumsen 2011). So-called first-generation biofuels, which are now often used, are edible. Second-generation biofuels are not edible, but they compete for land and other agricultural resources (such as fertilizer) with food crops and therefore also do not meet all requirements that derive from the norm to avoid increases in food prices. Currently, so-called third-generation biofuels are under development. These are based on bacteria and algae and are likely to meet the requirements with respect to the norm of not increasing food prices. However, they are still under development and expensive. Therefore they do not yet meet the design requirement of competitive price.

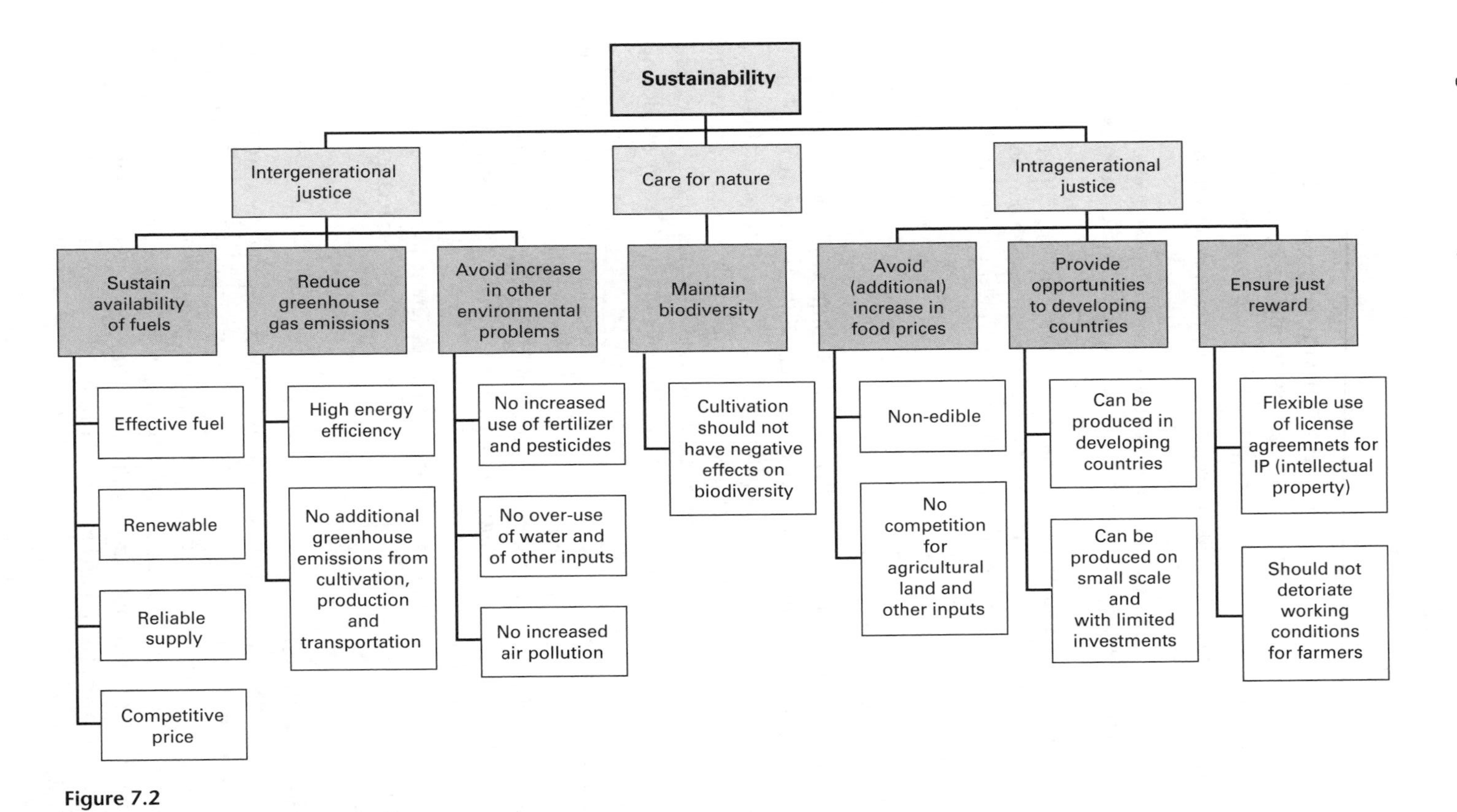

Figure 7.2
A possible values hierarchy for biofuels.

Value Conflict

I now want to look at how we might deal with value conflicts in design for sustainability taking the biofuels case as an illustrative example. A value conflict may be said to apply under the following conditions:

> 1. A choice has to be made between at least two options for which at least two values are relevant as choice criteria.
> 2. At least two different values select at least two different options as best. The reason for this condition is that if all values select the same option as the best one, we do not really face a value conflict.
> 3. The values do not trump each other. If one value trumps another any (small) amount of the first value is worth more than any (large) amount of the second value. If values trump each other, we can simply order the options with respect to the most important value and if two options score the same on this value we will examine the scores with respect to the second, less important, value. So if values trump each other, there is not a real value conflict. (van de Poel and Royakkers 2011, 177–178)

In the case of biofuels, we face a value conflict between intergenerational justice and intragenerational justice. While biofuels increase intergenerational justice by making fuels available to future generations and reducing emissions of greenhouse gases, they reduce intragenerational justice, in particular by competing with food production and so having an upward effect on food prices. From the viewpoint of intergenerational justice, we should probably choose first-generation or second-generation biofuels, and from the viewpoint of intragenerational justice we probably should prefer third-generation biofuels. The values of intergenerational and intragenerational justice are also not trumping each other because neither of them is overall more important than the other (I assume). So we have a value conflict.

I briefly want to discuss and assess three different strategies for dealing with this value conflict: life-cycle analysis (LCA), respecification, and innovation.

Life-Cycle Analysis

Life-cycle analysis is a tool with which to assess the environmental impacts of a product across its life cycle. It can be very useful in understanding design options, comparing them, and assessing whether they meet design requirements. To deal with value conflict, an LCA should not only assess the environmental impact on individual dimensions but should also provide an overall comparison of environmental impact. This is done by aggregating the environmental impacts into one measure (figure 7.3). To this end, weighing factors are given to the various environmental impacts to convert them into an overall measure. The UBP '06 method mentioned in figure 7.3, for example, weights the various environmental impacts by expressing them in "eco-points" (in German, Umweltbelastungspunkte, abbreviated UBP). The weighing of various impacts is based on sustainability targets set in the law (FOEN 2009).

The determination of weighing factors is an ethically and politically laden normative decision. In the case of UBP '06, the weighing factors are based on national laws, and thus the aggregated environmental impacts of various options may be different in different countries, depending on the scope and strictness of their laws. Other LCAs use other weighing factors, such as the Ecoindicator '99 mentioned in figure 7.3, which may lead to yet another ordering of the options.

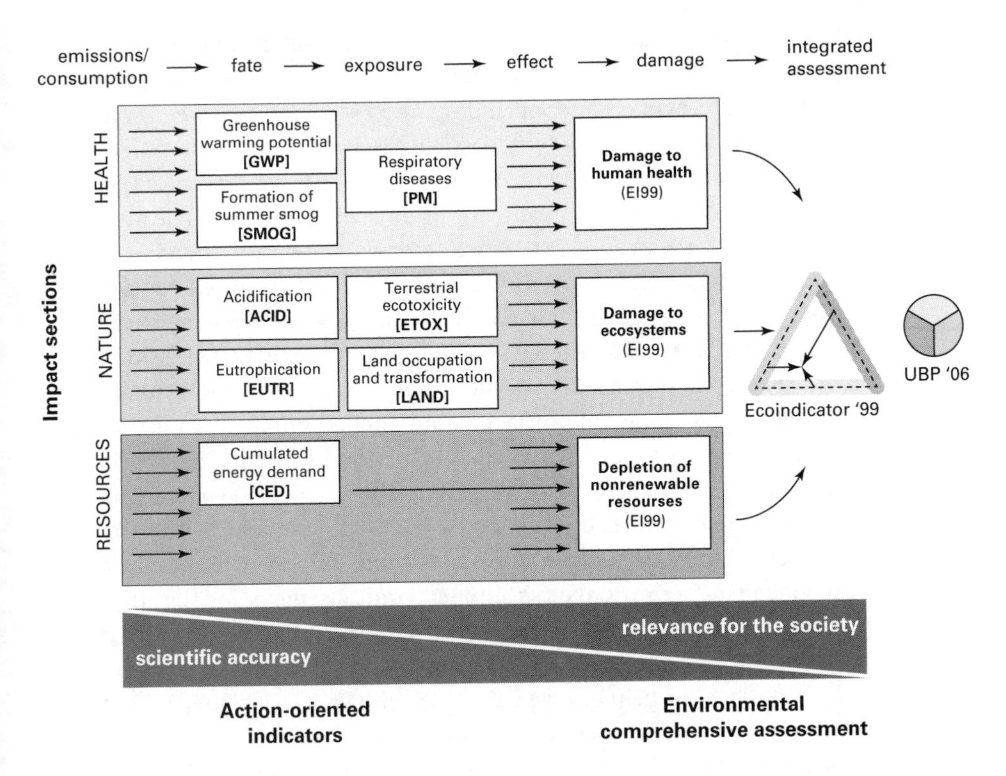

Figure 7.3
Life-cycle analysis and the construction of environmental impact indicators. Source: Zah et al. 2007.

There is, however, a more fundamental issue with the use of weighing factors in aggregating a wide diversity of environmental impacts into one measure. Aggregation supposes that it is always possible to compensate for a loss in one environmental dimension with a gain in another. But do lower emissions of greenhouse gases really compensate for a loss of biodiversity? Maybe we should aim at an option that meets some minimal standards with respect to both emissions of greenhouse gases and biodiversity rather than allow compensation for a loss in the one by a gain of another. Similarly, one might wonder whether a reduction in environmental impacts would compensate for an increase in hunger in the Third World, or the other way around. It has been suggested by philosophers, and others, that some values are incommensurable, in the sense that they cannot be expressed on a common scale and in the sense that a loss in one value cannot be fully compensated by a gain in another value (Raz 1986; Baron and Spranca 1997; Tetlock 2003).

Value incommensurability, then, seems to set a fundamental limit on the degree to which LCA can help to resolve value conflicts. For the same reason, the usefulness of other popular design tools to choose between conflicting options, such multiple criteria analysis (MCA) and cost–benefit analysis (CBA), is limited (Franssen 2005; Hansson 2007; van de Poel 2009). Like LCA, MCA and CBA also convert a wide range of value into single measure. This is most obvious in CBA, where money is used as the common denominator, but it is also apparent in MCA, which at least implicitly supposes that the various value dimensions that are used in the criteria are commensurable. The problem often is not only that weighted aggregation assumes value commensurability (which may be an untenable assumption), but also that the weighing factors are applied implicitly and thus normative assumptions remain implicit and are not discussed.

Let us, finally, look at an LCA of biofuels. Figure 7.4 shows a comparison of various available biofuels by greenhouse warming potential (GWP) and by two different measures for overall environmental impact. The latter take into account a range of various environmental impacts, including the effect on biodiversity, but they do not take into account such economic factors as costs and the potential effect on food prices. These measures of environmental impact cannot help resolve the value conflict, because they do not address the value of intragenerational justice.

Although the LCA of biofuels cannot resolve the value conflict, it is nevertheless interesting, and it is useful in design for sustainability. One thing we can learn from figure 7.4 is that, although most biofuels score better on GWP than traditional fuels, they score less well on other environmental

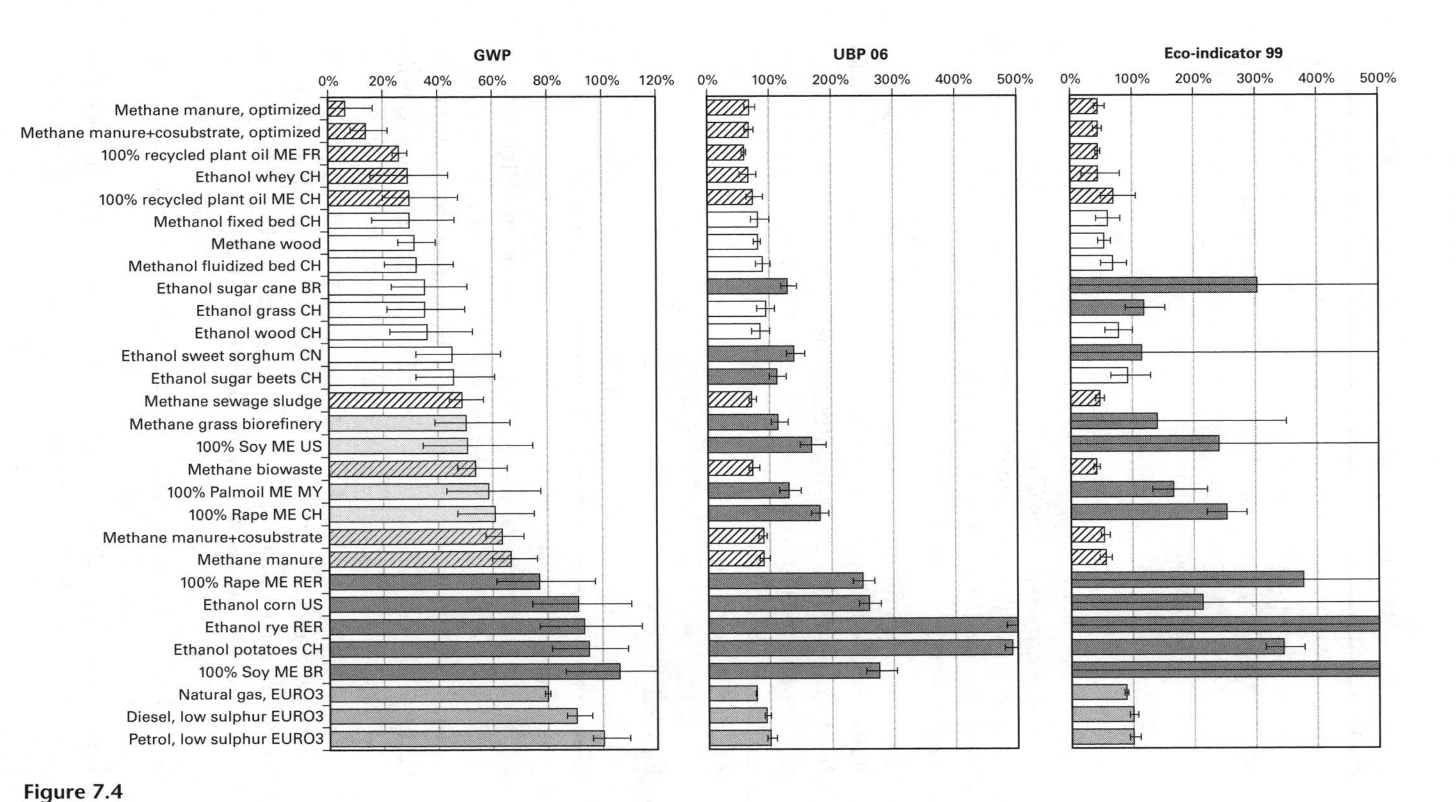

Figure 7.4
Overall environmental life-cycle assessment of all unblended biofuels studied in comparison to fossil reference. Source: Zah et al. 2007. (For color-coding, see original figure.)

dimension, even to such a degree that, according to the overall measures UBP '06 and Ecoindicator '99, for many of them the overall environmental impact is worse than that of traditional fuels. Of course, the latter assessment rests on certain assumptions about weighing factors that may be controversial. Nevertheless, it seems telling that UBP '06 and Ecoindicator '99 give similar results. But even apart from the overall comparison, the LCA points at same distinct environmental disadvantages of biofuels, which are summarized as follows on page xiv of Zah et al. 2007:

> [M]ost of the environmental impacts of biofuels are caused by agricultural cultivation. In the case of tropical agriculture this is primarily the slash-and-burning of rainforests which sets great quantities of CO_2 free, causes air pollution and has severe impacts on biodiversity. ... In the moderate latitudes it is partially the low crop yields, partially the intensive fertilizer use and mechanized tilling that cause the unfavorable environmental impacts.

We have seen that LCA has three main disadvantages as a way of dealing with value conflicts in design for sustainability: It may not cover all relevant values, it treats values as commensurable by applying weighing factors in aggregation, and the normative judgments that are applied in the weighing factors often remain implicit. Respecification may provide an interesting way to avoid some of these pitfalls

Respecification

By "respecification" I mean that a new specification is made of the higher-order values. In the case of biofuels, this would mean that the values of intergenerational justice, care for nature, and intragenerational justice are kept, but that the way these values are specified in terms of design requirements is changed from what is shown in figure 7.2. As we have seen, usually more than one specification is tenable. This means that we can often respecify the values in a values hierarchy and still have a justifiable specification. The trick now is to aim for a specification so that the value conflict is resolved or at least diminished. One might, for example, argue that the value of intergenerational justice would also allow biofuels that are more expensive than current fuels because this is a reasonable sacrifice to ask from the current generation. Alternatively, one could argue that intragenerational justice does not require that there be no competition for agricultural land and other inputs between biofuels and food crops, only that this competition be kept to a minimum.

Respecification may, as these examples already suggest, be controversial. People may disagree whether a certain respecification is indeed tenable.

Moreover, there is no guarantee that a respecification that resolves the value conflict is available. Even if we allow somewhat more expensive biofuels, no actual biofuels may meet all the new requirements. Nevertheless, respecification avoids some of the pitfalls of LCA, CBA, and MCA. First, if the initial values hierarchy contains all relevant values, it does not leave out crucial values. Second, it avoids the application of weighing factors and aggregation and does not treat values as directly commensurable. Third, it makes the normative value judgments that are made to resolve the value conflict explicit and open for discussion. Of course, this does not guarantee agreement, but it makes normative discussion possible, instead of proceeding on the basis of largely implicit normative assumptions.

Innovation

A third possible way to deal with value conflict is innovation (van den Hoven, Lokhorst, and van de Poel 2012). By "innovation" I mean the development of new options. Such new options might meet all the design requirements and so resolve the value conflict. In the case of biofuels, we can see how the value conflict already leads to innovation. A motive behind the development of so-called third-generation biofuels is the fact that current biofuels do not meet all the design criteria mentioned in figure 7.2. The case of biofuels, however, shows that innovation does not always resolve a value conflict. At present there are no biofuels that meet all the design requirements as we have seen. Still, ongoing R&D efforts may eventually lead to innovations so that all the requirements mentioned can be met.

Innovation may have also have disadvantages. It may, for example, lead to new products that have drawbacks. Third-generation biofuels may, for example, be based on genetic manipulation of algae, which may introduce new hazards. In the case of sustainability, the rebound effect is often cited as a possible disadvantage. If a solution is available without major direct disadvantages against comparable or lower costs, it may lead to additional consumption of the product; thus, if a good biofuels alternative is available for traditional fuels, this might frustrate attempts to reduce the fuel consumption of cars or even lead to higher consumption, so that the initial problem would not be solved and might sometimes even be exacerbated.

Value Conflict and Conceptions of Sustainability

Let us look briefly at how value conflict is related to the existence of various conceptions of sustainability. The first thing to note is that agreement on a conception of sustainability does not necessarily avoid value conflict. (See figure 7.2.) Although one might agree on a conception and a specification of sustainability in a particular case, that would not rule out the possibility that different design requirements would select different options. Nevertheless, there may be conceptions of sustainability that avoid value conflict. For example, if one believes that there is one metric for sustainability, in which all relevant value considerations can be expressed as is done in LCA, it may be possible to order all possible options on one scale and to avoid value conflict altogether.

Second, the existence of various conceptions of sustainability does not entail value conflict. As was argued above, there may be a shared specification of sustainability in a given case even if different actors have different conceptions of sustainability. This specification may, in combination with the available options, lead to a value conflict. But whether or not it does so does not depend on the existence, or on the absence, of different conceptions of sustainability.

Figure 7.5 summarizes the situation I have sketched in this chapter. On the most abstract and general level, we have a shared concept of sustainability. The agreement on that level, I have suggested, contains the idea that sustainability is desirable and the idea that sustainability is about the sustenance of some X into the future. On the second level, we have different conceptions of sustainability, which may specify X in quite different ways and which may refer to a range of more specific values, such as intergenerational justice, intragenerational justice, and care for nature. On the third level, we have context-dependent specifications of sustainability. As figure 7.5 shows, different specifications of sustainability in a specific case may cohere with one and the same conception of sustainability; on the other hand, some specifications of sustainability may cohere with different conceptions of sustainability.

Value conflicts usually are resolved at the level of specification, as in the strategy of respecification that I have discussed, or by the development of new options, as in the strategy of innovation. This means that agreement on a conception of sustainability is not necessary to deal with value conflicts in design for sustainability. Still, it seems reasonable to suspect that normative disagreement about the right conception of sustainability will make it harder to agree on a specification of sustainability and harder to resolve value conflicts. In some cases, it may be impossible to come to a shared specification of sustainability that coheres with the various conceptions of sustainability of the various actors involved.

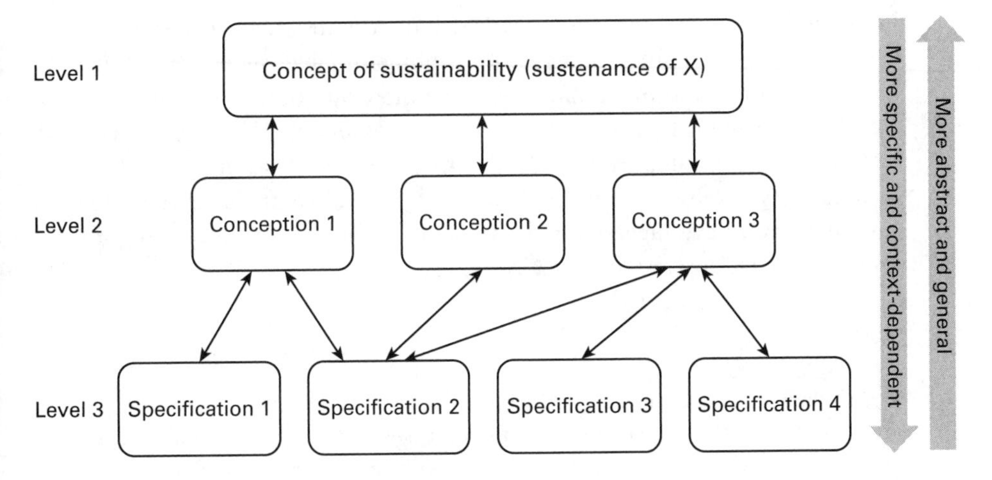

Figure 7.5
Relationship between concept, conceptions, and specifications of sustainability. In the case depicted, specification 2 is shared by the various conceptions.

An interesting way to deal with such cases has been suggested in the approach of Value Sensitive Design (VSD). VSD is an approach to the design of information systems that systemically takes into account values of ethical import (Friedman, Kahn, and Borning 2006). Two concepts that figure in VSD are "value dams" and "value flows" (Miller, Friedman, and Jancke 2007, 284). A value dam occurs when certain design features are strongly opposed by a stakeholder or by multiple stakeholders; a value flow occurs when a design feature is supported by a large number of stakeholders. Similarly, in design for sustainability, a value dam would occur with respect to design features strongly opposed by at least one relevant conception of sustainability, and a design flow would occur with respect to design requirements that fit a large number of conceptions of sustainability.

Conclusions

I have argued that sustainability is a contested concept, but that that fact does not frustrate the possibility of design for sustainability. In a concrete design project, sustainability always must be specified for that specific context, and I have offered the idea of a values hierarchy that can be used to make such specifications. Specifications of sustainability may raise value conflicts in the sense that none of the available options meets all the design requirements. I have suggested that life-cycle analysis, useful in design for sustainability, is often not the right instrument with which to resolve a value conflict underlain by a range of incommensurable values. Strategies of respecification and innovation might be more appropriate in such circumstances. Finally, I have discussed how the contestable character of sustainability might complicate dealing with values conflicts in design for sustainability as different stakeholders might have different conceptions of sustainability. I have suggested that "value dams" and "value flows" might provide a way to understand these complications.

Note

1. The Nuffield Council mentions three general values: human rights, solidarity, and environmental sustainability. The notion of sustainability used in the test is broader than environmental sustainability. Solidarity, as used by the Nuffield Council, is similar to what I call intragenerational justice. I treat human rights here as part of intragenerational justice.

References

Baron, Jonathan, and Mark Spranca. 1997. Protected values. *Organizational Behavior and Human Decision Processes* 70: 1–16.

Barry, Brian. 1999. Sustainability and intergenerational justice. In *Fairness and Futurity: Essays on Environmental Sustainability and Social Justice*, ed. Andrew Dobson. Oxford University Press.

Bhamra, Tracy, and Vicky Lofthouse. 2007. *Design for Sustainability: A Practical Approach*. Gower.

Birkeland, Janis. 2002. *Design for Sustainability: A Source Book for Ecological Integrated Solutions*. Earthscan.

Brumsen, Michiel. 2011. Sustainability, ethics and technology. In *Ethics, Technology, and Engineering*, ed. Ibo van de Poel and Lambèr Royakkers. Wiley-Blackwell.

Dobson, Andrew. 1998. *Justice and the Environment: Conceptions of Environmental Sustainability and Theories of Distributive Justice*. Oxford University Press.

FEANI. 2012. *FEANI position paper on Code of Conduct: Ethics and Conduct of Professional Engineers*. European Association of National engineering Societies 2006 [cited 11 June 2012]. Available from http://www.feani.org/site//index.php?eID=tx_nawsecuredl&u=0&file=fileadmin/PDF_Documents/Position_papers/Position_Paper_Code_of_Conduct_Ethics_approved_GA_2006.pdf&t=1339492458&hash=ed5883810769d5bfe5e05d4a8f33ce78578f81f9.

FOEN. 2009. The Ecological Scarcity Method—Eco-Factors 2006. A method for impact assessment in LCA. Bern: Federal Office for the Environment (FOEN).

Franssen, M. 2005. Arrow's theorem, multi-criteria decision problems and multi-attribute preferences in engineering design. *Research in Engineering Design* 16: 42–56.

Friedman, Batya, Peter H. Kahn Jr., and Alan Borning. 2006. Value sensitive design and information systems. In *Human-Computer Interaction in Management Information Systems: Foundations*, ed. Ping Zhang and Dennis Galletta. M. E. Sharpe.

Griffin, James. 1986. *Well-Being: Its Meaning, Measurement, and Moral Importance*. Clarendon.

Hansson, S. O. 2007. Philosophical problems in cost-benefit analysis. *Economics and Philosophy* 23: 163–183.

Inderwildi, Oliver R., and David A. King. 2009. Quo vadis biofuels? *Energy & Environmental Science* 2: 343–346.

Jacobs, Michael. 1999. Sustainable development as contested concept. In *Fairness and futurity: Essays on Environmental Sustainability and Social Justice*, ed. Andrew Dobson. Oxford University Press.

Miller, Jessica K., Batya Friedman, and Gavin Jancke. 2007. Value tensions in design: The value sensitive design, development, and appropriation of a corporation's groupware system. In *Proceedings of the 2007 International ACM Conference on Supporting Group Work*. ACM.

Naylor, Rosamond L., Adam J. Liska, Marshall B. Burke, Walter P. Falcon, Joanne C. Gaskell, Scott D. Rozelle, and Kenneth G. Cassman. 2007. The ripple effect: Biofuels, food security, and the environment. *Environment* 49: 30–43.

Norton, Bryan. 1999. Ecology and opportunity: Intergenerational equity and sustainable options. In *Fairness and Futurity: Essays on Environmental Sustainability AND Social Justice*, ed. Andrew Dobson. Oxford University Press.

NSPE. 2010. NSPE Code of Ethics for Engineers (http://www.nspe.org/Ethics/CodeofEthics/index.html).

Nuffield Council on Bioethics. 2011. Biofuels: Ethical Issues.

Nussbaum, M. C. 2000. *Women and Human Development: The Capabilities Approach*. Cambridge University Press.

Raz, Joseph. 1986. *The Morality of Freedom*. Oxford University Press.

Stitt, Fred A. 1999. *Ecological Design Handbook: Sustainable Strategies for Architecture, Landscape Architecture, Interior Design, AND Planning*. McGraw-Hill.

Tetlock, Philip E. 2003. Thinking the unthinkable: Sacred values and taboo cognitions. *Trends in Cognitive Sciences* 7: 320–324.

van den Hoven, Jeroen, Gert-Jan Lokhorst, and Ibo van de Poel. 2012. Engineering and the problem of moral overload. *Science and Engineering Ethics* 18: 143–155.

van de Poel, Ibo. 2009. Values in engineering design. In *Philosophy of Technology and Engineering Sciences*, volume 9, ed. Anthonie Meijers. Elsevier.

van de Poel, Ibo. 2012. Can we design for well-being? In *The Good Life in a Technological Age*, ed. Philip Brey, Adam Briggle, and Edward Spence. Routledge.

van de Poel, Ibo. 2013. Translating values into design requirements. In *Philosophy and Engineering: Reflections on Practice, Principles and Process*, ed. D. Mitchfelder, N. McCarty, and D. E. Goldberg. Springer.

van de Poel, Ibo, and Lambèr Royakkers. 2011. *Ethics, Technology and Engineering*. Wiley-Blackwell.

WCED. 1987. *Our Common Future: Report of the World Commission on Environment and Development*. Oxford University Press.

Zah, Rainer, Heinz Böni, Marcel Gauch, Roland Hischier, Martin Lehmann, and Patrick Wäger. 2007. Life Cycle Assessment of Energy Products: Environmental Assessment of Biofuels. Empa Technology and Society Lab.

8 Industrial Ecology and Environmental Design

Braden Allenby

Industrial ecology and its associated tools and methodologies are major means by which today's technocrats approach questions of environmental design. A discussion of their evolution and the institutions that were developed around them is interesting not just for the technical content, but also as a larger window on the development of environmental consciousness in the late twentieth century. But the fascination of this narrative should not be allowed to distract from a more basic initial question: What is "environmental design"?

There are five basic categories of environmental design, although other ways of framing such a developing area are certainly possible.

The most familiar may be using biological or ecological models or systems to guide or generate designs of materials, products, processes, facilities, or infrastructure. This practice is sometimes called "biomimicry" or "biomimetics" (Biomimicry 3.8 2012). It is not, however, new; remember that the Wright brothers were copying bird flight and dynamics when they worked on the design of their first airplanes.

The second is sometimes called "ecological engineering," and tends to focus on combining engineering with natural science to design ecosystems. The emphasis in ecological engineering is on the natural system rather than on built or human systems, and the goals often include biological conservation and development of "sustainable ecosystems."

The third category involves creating designs in which biological or ecological subsystems are incorporated into larger systems. An example of this category is the design of space colonies, or, perhaps more realistically in the short term, floating cities (NASA 1979). Many if not most existing physical human communities fall into this category but, of course, were not designed *de novo*.

The fourth category is design of integrated human, natural, and built systems at a regional and global scale, or "earth systems engineering and management" (Allenby 2007, 2011). Examples might include concerns about managing global water supplies, or about anthropogenic climate change, or about understanding and managing the phosphorous and nitrogen cycles. In this case, the understanding that humans are in fact engaged in terraforming their planet lags the obvious evidence of such terraforming, and the complexity of the systems involved and the unpredictability of human intervention in them are daunting. Nonetheless, that humans are impacting their planet is obvious, and eventually the need to take responsibility for those impacts will be accepted.

The fifth category, which does not neatly fit into any of the others, is the development of tools and methods for incorporating appropriate environmental considerations into all aspects of engineering design. This is the category that industrial ecology initially grew out of, but at this point industrial ecology studies and methods are broad enough to contribute data and analysis useful to other categories as well.

When one speaks of environmental design, then, it is not necessarily clear what is meant. Continuing to terraform the Earth, albeit perhaps more consciously, is a far cry from building a marsh to treat water from a farm before it enters a river—and both are a far cry from understanding the life-cycle environmental implications of an automobile or a computer chip. Industrial ecology studies and practices can certainly contribute to environmentally appropriate design at many scales, but is best thought of as a field that supports the fifth category identified above: environmentally appropriate design at the mesoscale.

What Is Industrial Ecology?

Industrial ecology is a field of study that has developed over the past 30 years as a response to increasing concern with environmental issues. It is an integrative endeavor, combining scientific and technological expertise and methodology with environmental science and policy. Because it is relatively new, and because its practitioners include both environmental activists and those trained in more scientific and engineering disciplines, the boundaries of the field are not yet well defined. As with any new domain, this is not just a technical issue; new ideas, approaches, concepts, and frameworks tend to arise in clumps, and jostle each other for preeminence. This was certainly true in the case of industrial ecology: Some concepts, such as "ecological economics" developed their own, complementary spheres of meaning, while others, such as "industrial metabolism" overlapped to a considerable degree with industrial ecology, and eventually were subsumed by it as a subfield (Ayers and Simonis 1994).

What matters is generally not the particular term used; "industrial ecology" could just as well be "industrial metabolism" with few operational implications. But the process of working out meaning, boundaries, and scope is important. It tends to define not just who is in (and out) of the relevant community, but what skills are required and what standards of truth and proof will be applied to work within the field. This is by no means a settled matter for industrial ecology. Thus, while some tools and methodologies, such as life-cycle assessment (LCA), Design for Environment (DfE), and mass flow analysis (MFA), are clearly central to industrial ecology, some are working on extending the field to include sustainability issues and metrics, raising issues that will be discussed below.

Some of the difference in perspective within industrial ecology can be illustrated by two commonly cited definitions of the field. The leading technical text on industrial ecology now defines it as follows, building on the definition provided in earlier editions:

> Industrial ecology is the means by which humanity can deliberately and rationally approach and maintain sustainability, given continued economic, cultural, and technological evolution. The concept requires that an industrial system be viewed not in isolation from its surrounding systems, but in concert with them. It is a systems view in which one seeks to optimize the total materials cycle from virgin material, to finished material, to component, to product, to obsolete product, and to ultimate disposal. Factors to be optimized include resources, energy, and capital. (Graedel and Allenby 2011, 32)

In contrast, an often-cited White Paper on Sustainable Development and Industrial Ecology issued by the Institute of Electrical and Electronic Engineers frames industrial ecology in more technical and less exhortatory terms:

> *Industrial ecology* is the objective, multidisciplinary study of industrial and economic systems and their linkages with fundamental natural systems. It incorporates, among other things, research involving energy supply and use, new materials, new technologies and technological systems, basic sciences, economics, law, management, and social sciences. Although still in the development stage, it provides the theoretical scientific basis upon which understanding, and reasoned improvement, of current practices can be based. … It is important to emphasize that industrial ecology is an objective field of study based on existing scientific and technological disciplines, not a form of industrial policy or planning system. (IEEE 1995, 1)

The two definitions have clear similarities. Each, for example, clearly understands industrial ecology as a systems-based and multidisciplinary approach, and is framed in technocratic terms. Each suggests that industrial ecology, while grounded in science and engineering, is broader in that it includes elements of law, economics, and the social sciences. Each also is clear that industrial ecology studies industrial and technological systems, differentiating industrial ecology from other environmental fields, which tend to emphasize biological and ecological systems and natural cycles rather than human systems. Both also reflect the roots of industrial ecology in environmentalism by mentioning traditional issues such as resources, energy, and capital; they neglect services, information and communication technologies, and the implications of emerging technologies (particularly the so-called Five Horsemen: nanotechnology, biotechnology, information and communication technology, robotics, and applied cognitive science) (Allenby 2011).

But there are also noteworthy and substantive differences. For example, the first definition begins by linking industrial ecology to "sustainability," while the second takes pains to claim industrial ecology as "an objective field of study." The first thus suggests a field with a normative bias, since the concept of "sustainability" involves, at the least, a privileged position for environmental values and an egalitarian political perspective. The second seeks to ground industrial ecology in the tradition of scientific objectivity, and is even at pains to claim identity with science rather than policy or planning. This very fundamental difference over whether the purpose of industrial ecology is to support environmental activism (or, more recently, the egalitarian perspectives characteristic of sustainability) or whether it is to provide scientific data and analyses without normative biases continues within the industrial ecology community.

There is a second fundamental trend in industrial ecology of which the reader should be aware. Industrial ecology as initially conceived began as a broadly environmental domain, as the biological analogy explicit in the term "industrial ecology" suggests. The first use of the term appears to have occurred in 1970 in the title of a short-lived environmental journal. In 1972, Japan's Ministry of Industry and International Trade (MITI) informally adopted the analogy between ecology and industrial systems as a model for the Japanese industrial system—an effort that faded with the more immediate policy issues raised by the 1973 energy crisis.

In retrospect, a more fundamental reason why industrial ecology failed to take hold in this early period of modern environmentalism was probably that environmentalists, government regulators, and industry were not yet sophisticated enough in their understanding of environmental issues to adopt a framework that treated these issues as inherent in industrial, technological, and economic systems, rather than as by-products of improper operations. Although it sounds trivial in retrospect, such an advance was a significant conceptual leap. After all, it was in the late 1970s and the 1980s that the complex, systemic nature of regional and global environmental perturbations such as ozone depletion, regional and global spread of pollutants, loss of biodiversity, climate change, and the inability of simplistic regulatory approaches to address them, became apparent. For example, it was not until 1992 that the international treaty regarding climate change, the United Nations Framework Convention on Climate Change, was negotiated.

It was also at this time that the implications of environmental liability began to change for industry. Regulatory costs began increasing dramatically; demands of stakeholders, customers, and regulators encouraged new, more systemic ways of thinking about environmental issues, in part to avoid unnecessary and steeply increasing costs; and the potential for unanticipated disruption of operations inherent in a shift from end-of-pipe to material, product, and factory management (e.g., banning chlorofluorocarbons by international agreement) became an important consideration. This last issue became quite apparent to industry professionals when, because of chlorofluorocarbons' impact on the stratospheric ozone layer, the international community moved to ban them via the Montreal Protocol (agreed in 1987; entered into force 1989). Whereas most previous forms of environmental regulation involved end-of-pipe controls and thus could be implemented without affecting manufacturing operations, CFCs were heavily used in manufacturing. Meeting this new generation of environmental regulations, then, was not simply a matter of buying a new scrubber or designing new water-treatment processes, but often required serious redesign of existing production and product systems to continue operations.

The successful introduction of industrial ecology thus had to wait until the late 1980s. In 1989, Robert Frosch and Nicholas Gallapoulos, both with General Motors, wrote an article for *Scientific American* suggesting that industrial systems could be more efficient if they were designed so that their material and energy flows mimicked those of biological ecosystems (Frosch and Gallapoulos 1989). This time, the concept gained immediate traction, especially at AT&T, which subsequently began development of specific design methodologies and recommendations such as Design for Environment (DfE) to provide engineering design tools to integrate environmental considerations and constraints into the product design process (Allenby 1991, 1992). AT&T turned to DfE because it could be introduced as an extension of an existing product realization process that was then standard in the electronics industry, known as "Design for X" or "DfX," where the "X" stood for a desired product characteristic such as testability, safety, manufacturability, or reliability (Gatenby and Foo 1990; Allenby 2011).

This path represented an understanding by the pioneers of industrial ecology that a fundamental change was best achieved by minimizing the appearance of radical change and blending with the status quo, thus minimizing the disruption, additional effort, and learning required of the target audience (in this case, product design teams). For similar reasons, the path of teaching design engineers about environmental issues in detail was avoided; rather, efforts were made to modify existing practices and tools so that environmental benefits would accrue without requiring any additional new procedures. For instance, significant benefits were achieved by revising an existing system of design manuals and parts-specification documents to avoid environmentally problematic choices. (For example, technically viable alternatives to mercury switches and PCB capacitors, which contained toxic materials, were recommended.)

Industrial ecology is of particular interest because its early adopters understood that they were developing a new domain that required both theoretical and institutional evolution. Thus, from the beginning there were also deliberate efforts to institutionalize industrial ecology as an industry practice. For example, AT&T led the formation of a Design for Environment Task Force at the American Electronics Association (AEA) in 1990 with the explicit goal of introducing industrial ecology practices and DfE to the electronics industry as a whole. The task force not only functioned as a multi-firm industrial ecology community, but also produced some important, albeit primitive, early materials—in particular, ten "AEA DfE White Papers" covering everything from materials recyclability to Design for Refurbishment to Design for Disassembly and Recyclability, and thus introduced the broader electronics industry to the concepts of DfE.[1] The first textbook on industrial ecology, *Industrial Ecology,* was written by two experts from AT&T with the deliberate intent of helping to ground industrial ecology in industrial and engineering domains, and thus to help further institutionalize it (and legitimize it as a field of study, and as an environmentally appropriate field, which was important because a number of traditional environmentalists, especially in EPA and many NGOs, regarded it as industry greenwashing) (Graedel and Allenby 1995).

For similar reasons, many of the firms and individuals active in the AEA effort were also active in establishing a set of annual conferences within the IEEE, the largest professional engineering organization in the world. The proceedings from this conference series, the IEEE International Symposium on Electronics and the Environment (subsequently renamed the International Symposium on Sustainable Systems and Technology), which began in 1993, are one of the best time series on DfE. A final initiative that was important in the early development of industrial ecology was the support of the National Academy of Engineering, which held a number of workshops on industrial ecology over this period, not just adding to the perceived legitimacy of the field but also publishing a number of edited volumes that established an intellectual foundation for further progress (Allenby and Richards 1994). The concluding initial steps in institutionalizing industrial ecology occurred with the founding of the International Society for Industrial Ecology, which opened to membership in February of 2001, and the establishment of a journal of record for the field, the *Journal of Industrial Ecology,* founded in 1997.

The specifics of the development of institutions and principles for industrial ecology should not distract from the broader context. Organic metaphors have long been popular, in part because they suggest a more environmental alternative to the mechanistic ones that they often displace. Conceptualizing the city as a mechanism, for example, was popular at one point, but has in recent decades been displaced by the metaphor of city as an ecosystem. Both are useful in some ways, but both are also simplistic and potentially misleading (Allenby 2011). Moreover, discussions of "green" industrial systems certainly predate the establishment of industrial ecology (Erkman 2002). Like most domains, then, industrial ecology should not be regarded as discontinuous or without precedent; rather, it is a codification and explication of many trends and ideas that, for a number of reasons, became viable at certain times. Moreover, industrial ecology strengthened, and was strengthened by, a number of complementary emerging concepts.

Ecological economics has already been mentioned. Another concept of note was that of life-cycle assessment, developed by the Society of Environmental Toxicology and Chemistry, founded in 1979. Indeed, although the initial activity regarding industrial ecology took place primarily in the United States, interest in systems-oriented and product-oriented environmental policies and activities subsequently shifted to Europe, where the interest in sustainability led to initiatives on product takeback and chemical regulation. These, in turn, encouraged the continuing development of methods and tools for industrial ecology. Presaging future trends in industrial ecology, the Netherlands took a more systemic approach to costs, benefits, and risks, moving beyond a purely environmental focus in an effort to define a sustainable society in one generation. The Netherlands publishing a National Environmental Policy Plan in 1989 and a National Environmental Policy Plan Plus in 1990 (VROM 1989, 1990). Somewhat awkwardly, industrial ecology remains primarily a European and American area of study, even though it continues to deal primarily with manufacturing facilities, products, and systems, most of which have moved to Asia.

Essentials of Industrial Ecology

The essence of industrial ecology is the biological analogy—more specifically the idea that ecosystems tend to use materials and energy efficiently, and that they evolved to do so. This leads to a simple Type I/Type II/Type III framework that, although biologically and historically inaccurate, is a useful explanation and framing of the underlying concept. Consider first a simple model of the first primitive soup, with a single form of organism that consumes what it can find in the ambient environment and excretes its wastes back into that environment. In such a Type I system, there is a linear flow: Food goes into a single organism, which consumes it and reproduces, and waste flows out into the environment. The environment's ability to supply sufficient food and its ability to absorb the waste are very simple factors limiting biomass growth in such a system.

The next step up in complexity is a Type II environment, where one has a number of different types of organisms: some compete for the original food, but others may learn to consume waste streams as their food, much as crops consume the manure that is used to fertilize their fields. In a Type II system, not only do definitions become more complex (whether a material is a waste or a food is no longer a characteristic of the material; now it is a characteristic of where it is within the system), but one can support a larger biomass than in a Type I system, all things equal. A Type III system, then, is one in which materials and energy are continually recycled within the biological community, only a small part of the material assets of the system becoming waste during any particular period. In this simple model, each level can support more biomass. One would anticipate that a Type III economy could support more activity than a Type I economy because of the internal cycling, all else equal.

This grossly simplistic schematic version of both biology and economic activity at least has the virtue of clarity. With a limited input of materials, and with limited sink capacity for wastes, it suggests that the way to achieve greater economic production is to increase the cycling of materials within the economy before a material or product is finally disposed of. But it also hides two important considerations. As regards energy, it is important to note that each jump in complexity will require more energy, all else equal. As a corollary, if an economic system is limited by either energy availability on the production side, or sink capacity on the energy-waste-disposal side (e.g., anthropogenic climate change and fossil fuels), the potential for internal cycling will be limited. Similarly, this model does not consider cost at all. If some waste-disposal options are not included in economic costs of production and consumption (that is, if they are externalities), it may be cheaper to dispose of wastes than to transport them and transform them for further economic use.

In addition to this simple conceptual model, industrial ecology has several basic themes. Among the most important are the following:

- Industrial ecology analyzes systems, rather than artifacts or specific natural components elements of the system. This is one reason why life-cycle analysis, which looks at a material or product over its life cycle rather than simply at a single state such as manufacture, use, or disposal, is such an important tool for industrial ecology.
- As part of this systemic approach, industrial ecology tends to emphasize long term concerns that are regional or global in scope and are persistent and difficult to manage, rather than particular incidents or events.
- Industrial ecology tends toward a scientific and engineering perspective, as opposed to, for example, ecological engineering, which tends to emphasize conservation biology and ecology.
- Combining the engineering and systems perspectives, an important focus of many industrial ecology studies is the resiliency of the integrated human/natural/built systems with which industrial ecology engages. Resilience is a very important, and very complex, property of complex adaptive systems. It varies over time as the underlying system does, making this area of industrial ecology both intellectually challenging and satisfying.
- An important thematic difference between industrial ecology and other environmental fields is the engagement with private firms and industrial systems at multiple scales. Industrial ecology sees this engagement as critical to managing environmental impact.
- Scale issues are very important to industrial ecology. Different studies may be carried out at many different scales. An LCA study might analyze a particular material, process, or product; an integrated industrial ecology study

might evaluate a facility or set of facilities (for example, ecopark studies); an MFA study might concentrate on material flows at broader regional, national, or global scale. Different issues will arise in each case, and the appropriate level for analysis of a particular issue may vary as well.

Development of DfE

Given industrial ecology's technocratic focus on product, process, and system design, it is useful to consider briefly the challenges that come with any engineering design process. Especially when a designer or a design team begins operating in a relatively undefined, somewhat ideological, and frequently highly charged area such as the environment, it is important that the practitioner remain grounded. The function of engineering in general, and design in particular, is, after all, to solve a problem. The problem may take many specific guises, but it almost always requires the creation of components, products, processes, infrastructures, practices, and services that meet social wants and needs in ways that are efficient, practical, manufacturable, economic, competitive, and timely—and, of course, give due consideration to relevant environmental concerns. And this activity is fairly stringently bounded: Whether working for a private firm, a governmental entity, or as a consultant, a designer's primary responsibility is to produce a rational and responsible solution that works in the real world. Increasingly this means that the designer must include the preferences of activist stakeholders, who may have very strong worldviews and very specific concerns, especially regarding environmental issues or human rights, but who also can mobilize public attention to demand attention. Globalization, with its mix of differing cultures, technological infrastructures, and regulatory regimes, adds to the complexity.

This is the domain of industrial ecology. Indeed, part of the technocratic aspect of industrial ecology arises because designers, if they are to achieve solutions in such complicated spaces, often using very complicated technologies, rely on quantitative methods both by inclination and by methodological choice. Their design solutions are created by using algorithmic treatments of design objectives and constraints—which, these days, almost always include some aspects of environmental concern. Unlike a policy maker or an activist, a designer cannot retreat to the luxury of constructive ambiguity, because at the end of the process they must have a real, workable, and competitive product or service.

It is therefore not surprising that industrial ecology from the beginning focused on design, and thus began working on DfE techniques. Note that

DfE is not the same as LCA. A life-cycle assessment provides data regarding a particular material in a particular use, or a certain product, but it does not provide a mechanism by which such data can be incorporated into better practices or design. DfE, on the other hand, has always been intended to be a tool that is effective in supporting better designs. That was one reason why both the name of the method, and the mechanisms that it adopted, were selected. DfE was positioned from the beginning to be a component of an existing product realization process that was already in wide use. Only later did the scope of DfE expand to include a generic set of practices applying industrial ecology principles to the design of such varied and complex products as consumer electronics, automobile subassemblies, and aircraft.

Even though DfE deals directly with environmental issues rather than the far more complex sustainability issues, it is still very difficult to balance different risks and create systemic mechanisms. While there are increasing numbers of large commercial software packages available to conduct LCA and, to a lesser extent, DfE analyses, a few of the more common heuristics, based at this point on several decades of experience, will give the non-engineer a sense of the sort of areas of interest that DfE covers. Some common heuristics for product design and manufacturing include the following:

- Design in manufacturing so as to minimize energy, material, and toxics use, and the number of separate processes, all else equal.
- Where possible, every molecule going in to a production process should come out in a product or commercially useful by-product, all else equal.
- Include environmental stakeholder values in design objectives and constraints (this is a separate requirement from simply including environmental considerations, since some stakeholders may have strong preferences for certain environmental goods as opposed to others that are ideological rather than scientifically supportable or justified).
- Design manufactured products so as to reduce energy and material consumption over product life cycle.
- Reduce packaging, including not only the packaging used for products that are produced and sold but also the packaging used for subassemblies or other pieces that are used in constructing the product, all else equal.
- Reduce use of toxics in manufacturing processes, and reduce presence of toxics in manufactured products, with particular attention paid to toxics that might leach or otherwise result in human or ecosystem exposure, all else equal.

• Remember that no product exists in a vacuum, so design of products should take account of their roles in services, networks and infrastructure, and social and cultural patterns, all else equal.
• Design for long product life, all else equal.
• Design to allow for environmentally appropriate adaptation to technological evolution and concomitant efficiency gains (so, for example, where a building is being put up in an area conducive to solar power production, orient the building so that installation of such technology will be cheap and easy; or site a new fossil-fuel-burning power plant in such a way that carbon capture from the stacks can economically be installed), all else equal.

The reader will have noted the constant refrain of "all else equal." This is important because usually all else is not equal, so tradeoffs must be made. For example, aqueous cleaning solutions were a common substitute for CFC cleaners when ozone depletion was recognized as a major environmental issue; such technologies do not release ozone depleting substances, but they do require more energy to operate because drying water off a part takes more energy than drying CFCs off a part. Thus, the tradeoff was increased emission of greenhouse gases for reduced emissions of CFCs. Under the circumstances, this was probably a good decision, but it does illustrate the need to avoid doctrinaire positions in activities as complicated as design. Moreover, some heuristics must be applied with great care. Thus, for example, designing for long product life may be a good idea generally, but if one is in a field where product functionality is changing rapidly, this can lead to expensive and non-competitive designs—cell phones that lasted for many years, for example, would be a poor design because many people trade them in for enhanced functionality and flexibility in only a year or two. Similarly, designing 1960s muscle cars so that they would have very long lives would have locked in substantial tailpipe emissions, rather than enabling a relatively rapid changeover to much more environmentally efficient automobiles as that technology was developed and rolled out to market.

Strengths and Weaknesses of Industrial Ecology

The history and the origins of industrial ecology as a concept, as a technocratic practice, and as a set of tools and methods go far to explain some of its more obvious strengths, weaknesses, and characteristics.

To begin with the point noted above, the development and implementation of industrial ecology has not reflected adequately the changes and shifts in either environmentalism as a social construct, or the globalization of the economy. Thus, a major trend in environmentalism has been the shift over the past several decades from American waste-reduction and remediation policies to European product take-back and life-cycle product-management initiatives—a shift that has if anything been even stronger in the case of sustainability, which has been far more popular conceptually and operationally in the European Union than in the United States or East Asia. While Japanese experts in industrial ecology have participated in the community since its beginning, and Japan has generally had an active DfE community, China and other Asian manufacturing powers have been less engaged. This has become especially notable as China has become the primary manufacturing power in the world, displacing the US. What this means is that the industrial ecology dialog about environmentally appropriate manufacturing and industrial systems has become somewhat disconnected from the societies, and the economies, that are actually doing the manufacturing. In reality, anyone who is interested in industrial ecology needs to at least be familiar with developments in Europe in order to keep up with regulatory initiatives and theory, and Asia to understand what is actually happening in practice.

Additionally, any discipline or area of practice that focuses on environmental design is seriously constrained by the ideological framework that environmentalism brings along with it. To be sure, it is important in our age, when it is becoming obvious that the Earth is increasingly impacted across all systems and scales by human activity, to consider the environmental implications of most activities, especially design activities. Not to do so in what scientists are increasingly calling the Anthropocene—the age of the human—would be irresponsible. But at the same time, this means that environmentalism is not the only value that must be accommodated in design—traditional issues such as time to market, manufacturability, consumer acceptance, health and safety of employees, material availability, and cost do not simply evaporate. Moreover, many of the systems with which industrial ecology grapples are technically complex adaptive systems and in such cases no single ideology, no matter how desirable in theory, can provide more than a partial perspective. This is one reason, among many, that the UN effort on climate change generally, and the Kyoto Protocol in particular, have faltered. Accordingly, it is important to understand the limitations, as well as the necessity, of an environmental approach.

In particular, it is important to question whether the fact that industrial ecology and many of its methodologies (e.g., DfE and LCA) as currently practiced remain focused to a large extent on environmental issues is a strength, or, increasingly, a weakness. For example, the original insight into the importance of conceptualizing a product through its life cycle to be able to understand and enable management of environmental considerations was a significant step forward. But it may be time for the next step, to sustainable engineering, to replace a simple environmental approach. This would fit with the way industrial ecology evolved initially, when it reflected the shift away from a relatively unsophisticated understanding of environmental issues, and toward a more comprehensive understanding of the anthropogenic Earth.

One clear trend in the development of industrial ecology has been a disproportionate focus on manufacturing, material and energy flow, and manufactured products. While some attention has been paid to services and to non-manufacturing sectors, very little of industrial ecology has focused on information and communication technologies and services, or the implications of information flows and networks. In part, this is simply another reflection of the history of the field: People involved in industrial ecology, especially in its early phases, came from environmental backgrounds, and much of the activism and regulation that characterized this period were focused on manufacturing sectors, issues, and characteristics (e.g., material flows). Nonetheless, in an era when technology systems are evolving rapidly and unpredictably across the entire technological frontier, this trend may be understandable, but it is also a serious weakness.

A more subtle related feature of industrial ecology is its grounding in an industrial, and thus technocratic, zeitgeist. Especially at the early formative stage, it was developed primarily by people trained in science and engineering, working in industry, and especially in manufacturing industries. Indeed, the idea of an ecology, which is a system of material and energy transfers and conversions, lends itself to manufacturing and industrial activities (as opposed to, say, information flows which might suggest a computational or neural analogy). In fact, because industrial ecology was primarily developed by industry experts, and supported by their firms, many environmentalists initially strongly opposed it, claiming that it was merely another industry effort to co-opt activist environmentalism. As industrial ecology has matured, the concern has lessened, but the implicit tension between activism, and the technologists' emphasis on data and objective analysis, remains.

Additionally, the historical background of industrial ecology positions it as a discourse arising in the context of a high-technology developed economy. Developed in industrial sectors in advanced economies, and as a response to environmental problems as perceived, prioritized, and valued in such countries, industrial ecology has not yet been particularly effective in the context of developing economies. This is not necessarily a strong weakness in purely environmental arenas, since environmental effects can generally be measured and quantified independent of social context, but it is clearly a strong weakness if one extends industrial ecology to sustainability dimensions, which bring in social and cultural domains that are perceived quite differently depending on economic development status. There are far more industrial ecology studies of artifacts such as automobiles and computer chips than there are of water quality, or sewage management, in poor regions of the world.

But possibly the most subtle issues arise from the implications of the term industrial ecology itself. Particularly because of the environmental sympathies of many sustainability and industrial ecology practitioners, there appears to be a strong tendency to accept the implicit analogy between biological systems and human systems far too literally. While there are certainly useful design ideas for both products and materials derived from study of "natural" systems, there is no necessary guarantee that just because something is natural or renewable it is therefore preferable. For example, one of the substitutes for CFCs in industrial cleaning processes was a terpene-bases system; proponents claimed it was safe because the active ingredient was also found in many plants, including pines. The rub was that as an essential oil it occurred in small concentrations, whereas in industrial settings it was often used in 70 percent or 80 percent solutions; the health and safety implications and the environmental implications of the latter are potentially far different than in the case of the former. Many toxins are, of course, entirely natural; this does not mean they are therefore preferable to designed materials in product or production systems.

Moreover, there was (and is) a tendency to mistake the value of the ecological metaphor—city or product or factory as ecosystem—as a statement about the fundamental behavior, complexity, and essence of a human system. In this, the term misleads: Human systems, whether city or corporation or community, are far more complex than natural biological or ecological systems in very important ways. Most obviously, human systems display a level of reflexivity, and behavior, and interaction with both cultural and technology systems, that natural systems lack. This level of complexity is sometimes called "wicked complexity" (Allenby and Sarewitz 2011; Rittel and Webber 1973). Unless industrial ecology, industrial metabolism, and similar terms are interpreted with a certain amount of sophistication, and the particular challenges of each design identified and evaluated rationally, over-reliance on biological or ecological metaphor can be as inappropriate as any other analytical deficiency.

Conclusion

Industrial ecology is of interest not just because of the suite of methods and tools that it encompasses, which enable design of environmentally preferable processes, products, facilities, and systems while meeting other relevant design objectives and constraints. The field also provides an interesting example of a deliberately created and institutionalized discourse, one that continues to reflect both its past, and the ideological, cultural, and policy tensions that have characterized the evolution of the environmental movement. It cannot yet be regarded as mature, if only because the cultural and technical context within which it is embedded—environmental science, technology, theory, and ideology—is continuing its rapid pace of change. Thus, the new textbooks on industrial ecology begin to explore the extension of the field into sustainable engineering—a profound challenge to practitioners and theorists alike (Allenby 2011).

Note

1. The ten White Papers were "What Is Design for Environment?" (B. Allenby, AT&T); "DfE and Pollution Prevention" (B. Allenby, AT&T); "Design for Disassembly and Recyclability" (R. Grossman, IBM); "Design for Environmentally Sound Processing" (J. Sekutowski, AT&T); "Design for Materials Recyclability (W. Rosenberg, COMPAQ, and B. Terry, Pitney-Bowes); "Cultural and Organizational Issues Related to DfE" (B. Allenby, AT&T); "Design for Maintainability" (E. Morehouse Jr., US Air Force); "Design for Environmentally Responsible Packaging" (K. Rasmussen, General Electric); "Design for Refurbishment" (J. Azar, Xerox); and "Sustainable Development, Industrial Ecology, and Design for Environment" (B. Allenby, AT&T).

References

Allenby, B. R. 1991. Design for environment: A tool whose time has come. *SSA Journal*, September: 5–9.

Allenby, B. R. 1992. Design for Environment: Implementing Industrial Ecology. PhD thesis, University of Michigan. University Microfilms pub. no. 9232896.

Allenby, B. R. 2007. Earth systems engineering and management: A manifesto. *Environmental Science & Technology* 41 (23): 7960–7966.

Allenby, B. R. 2011. *The Theory and Practice of Sustainable Engineering*. Prentice-Hall.

Allenby, B. R., and D. J. Richards. 1994. *The Greening of Industrial Ecosystems*. National Academy Press.

Allenby, B. R., and Daniel Sarewitz. 2011. *The Techno-Human Condition*. MIT Press.

Ayres, R. U., and U. E. Simonis, eds. 1994. *Industrial Metabolism*. United Nations University Press.

Biomimicry 3.8. 2013. biomimicry.net, accessed January, 2013.

Erkman, S. 2002. The recent history of industrial ecology. In *A Handbook of Industrial Ecology*, ed. R. U. Ayres and L. W. Ayres. Elgar.

Frosch, R. A., and N. E. Gallopoulos. 1989. Strategies for manufacturing. *Scientific American* 261 (3): 94–102.

Gatenby, D. A., and G. Foo. 1990. Design for X: Key to competitive, profitable markets. *AT&T Technical Journal* 63 (3): 2–13.

Graedel, T. E., and B. R. Allenby. 1995. *Industrial Ecology*. Prentice-Hall.

Graedel, T., and B. R. Allenby. 2011. *Industrial Ecology and Sustainable Engineering*. Prentice-Hall.

IEEE. 1995. White Paper on Sustainable Development and Industrial Ecology).

NASA (National Aeronautics and Space Administration). 1979. Space Resources and Space Settlement (NASA SP-428).

Rittel, H., and M. Webber. 1973. Dilemmas in a general theory of planning. *Policy Sciences* 4: 155–169.

VROM. (Netherlands Ministry of Housing, Physical Planning and the Environment). 1989. National Environmental Policy Plan (To Choose or to Lose).

VROM. 1990. National Environmental Policy Plan Plus. VROM 00278/10–90.

9 Ecodesign in the Era of Symbolic Consumption

Zhang Wei

In response to environmental problems, ecodesign is a new paradigm in the field of design focused on environmentally sustainable technologies. It has become an important topic not only in design circles, but also in environmental ethics, economics, management, and other fields. Sim Van der Ryn and Stuart Cowan (1996) claim that the environmental crisis is, in a sense, the crisis of design. So an appropriate choice to solve environmental issues might be though design. Indeed the implementation of ecodesign has made great strides in the past 20 years, although it continues to face several challenges.

The main challenge is that ecodesign pays too much attention to the environmental effects of the products themselves but seldom cares for the possible impacts on the behaviors of the users. That may lead to the consequence of the Jevons paradox—i.e., in some cases, the undesirable effects may offset the environmental merit of the products, which is called "backfire." For example, energy-saving lights are designed with the original intention of saving power, but for this very reason they are installed in the places where there was no light before, such as a garden or a garage. Contrary to the original intention, more electricity is then consumed (Verbeek 2005). Similarly, reducing the energy consumption of vending machines leads to their proliferation in more locations. And more efficient air travel has generated low-price airlines and therefore increased the number of people flying.

This problem is aggravated in the era of symbolic consumption. In this era, we purchase goods not only for their uses but also for their symbolic meaning. In other words, we consume goods not only to create and sustain the self, but also to locate us in society (Elliott 1994). Many products that are discarded are still functionally usable but no longer satisfy psychological needs or desires. "As long as this happens, it does not matter how 'clean' a product is. It is at least as important to try to get people not to throw away their things so quickly." (Verbeek 2005) This means that designers should not only pay attention to the physical durability of products but should also consider their psychological durability in the design process.

This chapter begins by characterizing the current problem of ecodesign by reflecting its basic rationale and methods. It then turns to the characteristics of symbolic consumption and its impacts on ecodesign. Finally, this chapter proposes a both/and approach to ecodesign as a response to the challenges of the Jevons paradox and symbolic consumption. This approach will not only focus on minimizing the environmental impacts of the products themselves, but will also pay attention to the interaction between the products and the users. In this approach, "the dichotomy between people and things is overcome precisely by taking into consideration their mutual relations" (Verbeek 2005). Several specific methods, including Design with Intent, User-Centered Design, and Emotional Design, will be introduced.

The Problems of Current Ecodesign

Ecodesign is an important advance because the traditional design paradigm does not take environmental sustainability and ecological impacts into consideration. Energy, for example, is usually nonrenewable, relying on fossil fuels and consuming natural resources. High-quality materials are used inefficiently, and as a result toxic and low-quality materials are discarded in soil, air, and water. To a large extent, ecological crises are related to such unsustainable design.

Under pressure from environmentalist movements, companies are challenged to contribute to sustainability by improving the environmental performance of their current forms of production, reducing the amounts of natural resources, energy consumption, and toxic emissions related to manufacturing, and reducing the use and disposal of their products and services (van der Zwan and Bhamra 2003). The term "ecodesign" refers to "any form of design that minimizes environmentally destructive impacts by integrating itself with living processes" (Van der Ryn and Cowan 1996). Ecodesign is not just a subdiscipline of design, as industrial design, landscape, and urban design are; it is a form of engagement and partnership with nature. The principles and methods of ecodesign can be applied to any field of design, such as the designing of spaces, buildings, cities, or products. I shall restrict myself to product design in this chapter.

In product design, the aim of ecodesign is to reduce and balance the adverse impacts of a manufactured product on the environment by considering the product's whole life cycle, from acquisition of raw materials through manufacturing, distribution, and use to recycling and disposal (Bhamra 2004). One model of ecodesign, employed by a number of organizations and governments, is the *Cradle to Cradle* model, developed by William McDonough (2002). McDonough claimed that the traditional cradle-to-grave manufacturing model wastes as much as 90 percent of the materials it uses, many of them toxic. On the Cradle to Cradle model, all materials are divided into two categories: "technical nutrients" and "biological nutrients." "Technical nutrients" are non-toxic, non-harmful inorganic or synthetic materials that have no negative effects on the environment and can be used in multiple cycles as the same product without losing their integrity or quality. "Biological nutrients" are organic materials that can be disposed of in any natural environment and that decompose into the soil after use (ibid.).

The main methods usually employed in ecodesign are minimizing the use of resources and energy during manufacture, adopting renewable resources and energy, improving energy efficiency, reducing emissions of wastes and toxic substances, and making a product physically durable, reusable, and recyclable. This strategy is, of course, problematic when, for example, we buy energy-saving lights but then leave them on all night (Lockton, Harrison, and Stanton 2009), or when we install lights in places where there is no need for them. Then the overall consumption of electricity increases rather than decreases. The result suggests that environmental impact is decided not only by the product but also by how people use it.

If ecodesigners ignore the potential influences of the product on the users in the design process, a Jevons effect may result. In *The Coal Question*, William Jevons (2010) observes that, contrary to our intuition, technological improvements in the efficiency of coal use increase consumption. He concludes that technological progress that increases the efficiency with which a resource is used tends to increase, rather than decrease, the rate of consumption of that resource. We can understand this paradox as follows: When the efficiency of resource use increases, if the market price of this resource remains constant, then its real price decreases when measured in terms of what it can achieve. With the decrease in the resource's price, the quantity of demand will increase, which is known as "rebound effect." When the rebound effect exceeds the original increase in efficiency from the technology progress, "backfire" occurs. Thus, the efficiency approach to sustainability adopted by ecodesign is dubious.

Symbolic Consumption and Its Influences on Ecodesign

Apart from the Jevons paradox, another phenomenon that also exhibits the limits of current ecodesign is symbolic consumption, a characteristic of advanced industrialized societies (Baudrillard 1998). Goods not only have a use value and an exchange value, as traditional economics believes; they also have a "sign exchange value" that indicates distinction, taste, and social status. For example, a BMW automobile certainly has both use and exchange value, but we also can understand its status as a sign in the code of consumer values: It signifies social distinction. For example, a coffee-maker is a means not only for producing coffee but also for exhibiting one's taste (Verbeek 2005).

With improvements in the overall standard of living, people have freed themselves from the era of material scarcity, in which the focus was on the physical functionality of the products. As long as things are workable they will not be discarded. But in the era of symbolic consumption, people pay more attention to the symbolic meanings of products, which play the leading role in consumer choice. Most of us have had the experience in a store of encountering two nearly identical goods whose prices differ considerably. Some people would choose the more expensive item just because it bears a famous brand name and thus carries more symbolic meaning than the other item.

According to Baudrillard, the symbolic meaning of a commodity is not invariable but rather is a result of social construction, mainly by advertisements. The media are such a powerful force in modern life that our aesthetic preferences, our values, and even our worldviews are strongly influenced by them. Every day we are bombarded with messages from the Internet, television, billboards, and other media. These messages promote not only products but also moods, attitudes, and a sense of what is and what is not important. Baudrillard (1998) claims that in the media age we experience "the death of the real." We actually live our lives in the realm of hyper-reality, connecting more and more deeply to things that merely simulate reality (e.g., television shows, music videos, and video games). The result is that these simulated images become more real to us than the physical reality that surrounds us (ibid.).

When the emphasis on consumption moves from physical function to symbolic meaning, and the traditional strategy of "physical obsolescence" is no longer an effective way for producers to increase their profits, the tricky producers create a new strategy: "psychological obsolescence" (Chapman 2009). They regularly change the styling of their products, then launch advertising campaigns to try to convince customers to update their stuff more frequently. This phenomenon is especially clear in the field of domestic electronic production. Every year, billions of domestic electronic products, including laptop computers, MP3 players, cameras, and cell phones, are discarded. Most of them are still functionally workable but are no longer *psychological* workable.

Apart from advertising, another reason, for the weakening of the emotional bond between users and things—a reason cited by Albert Borgmann (1984)—is that consumption diminishes the engagement of human beings with the products. For example, in the past people used to warm their houses by collecting firewood and stoking their fireplaces, but now all they have to do is turn up a thermostat. The task of heating is delegated to hidden machinery. We can enjoy commodities without having to engage ourselves with processes (ibid.). With the decrease in engagement that the thermostat brings, people no longer have an emotional bond with a thermostat as strong as the one they once had with a fireplace, and they don't see many differences between one model of thermostat and another; they are interchangeable and replaceable.

These reasons point another shortcoming of the current approach of ecodesign: It does not seek to prevent or postpone the throwing away of products, but rather to minimize the amount of environmental damage created by their disposal. According to Verbeek (2005), the most relevant issue for durable product development is "to devise ways to lengthen this psychological lifetime." No matter how environmentally friendly a product itself is, if designers still focus exclusively on its physical function and ignore the psychological needs or desires of users, the product will not escape the fate of being abandoned and all of the elaborate efforts of the designer will vanish like soap bubbles. This warns us once again that, in order to deal with environmental problems adequately, we should take an integrated approach in which "the relation between humans and products takes center stage, instead of focusing separately on humans (who should act more friendly toward the environment) and things (which should be 'cleaner')" (Verbeek 2005). I call it a *both/and approach.*

A Both/And Approach to Ecodesign

These two challenges of current ecodesign—the Jevons paradox and the weakening of the emotional bond between people and products—have several possible solutions. One is to control the price of the resource—for example, by levying a tax to suppress energy consumption when the efficiency of energy use is improved. Another is to help consumers develop sustainable lifestyles by means of environmental education. A third is to formulate environmental policies to regulate consumption, thereby reducing the amount of waste. But all these strategies rely on the use of external power to force the consumers to perform sustainable behaviors. Can we encourage people to spontaneously perform sustainable behaviors? Maybe a both/and approach to ecodesign could solve this problem.

"Both/and" means that both the environmental properties of the products and the influences on consumers' behavior are considered in the design process. The methods of the "both" part have already been solved by ecodesign. In this section, I will address the methods of the "and" part—that is, how to encourage sustainable behaviors through design, and how to increase the emotional bond between people and products. I will discuss three methods: *Design with Intent, User-Centered Design,* and *Emotional Design.*

Design with Intent

Design with Intent is an interdisciplinary approach that draws on the experiences of many design fields. It is defined as "design intended to influence or result in certain user behavior" (Lockton et al. 2009).The basic idea is "to modify or redesign the system to influence users' behavior towards a particular 'target behavior'" (Lockton et al. 2010). In their Design with Intent Toolkit, Don Lockton and his colleagues provide us with many practical methods. The toolkit includes many specific patterns that are grouped into six "lenses," each presenting a different perspective on design and behavior.

The six lenses are Architectural, Error-proofing, Persuasive, Visual, Cognitive, and Security. The architectural patterns include positioning and layout, material properties, segmentation, and spacing. For example, the layouts of supermarkets, shopping malls, and offices can influence the paths taken by users, exposing them to shelves, shops, and colleagues in a strategic order or hierarchy. Although these patterns are usually used in architecture and urban planning, they can also be applied to product design and even in software. Errorproofing patterns include defaults, interlock, extra step, and specialized affordances. For example, one can design "good" default settings and options (since many users stick with defaults) and change them only when one feels that it is really necessary to do so.

Persuasive patterns includes self-monitoring, reduction, tailoring, tunneling, and feedback from form. For example, self-monitoring can involve real-time feedback on the consequence of different behaviors, so that the correct next step can be taken immediately. Visual patterns include prominence and visibility, metaphors, perceived affordances, and implied sequences. For example, one can design certain elements so they are more prominent, obvious, memorable, or visible than others, so as to direct users' attention toward them, thereby making it easier for users to pick up the intended message or to pick the best options. Cognitive patterns include social proofing, farming, reciprocation, commitment, and consistency. Security patterns include surveillance, atmospherics, and threat of damage. For example, users will often decide what to do on the basis of what those around them do, or how popular an option is, and designers can make strategic use of this fact to influence behaviors (Lockton et al. 2009, 2010).

Apart from these practical design methods, the Design with Intent Toolkit also provides the designers a clear classification of target behaviors. With these standards, the success of such a design can be measured. It also should be noted that Design with Intent was developed primarily in response to a need to influence users' behavior so as to reduce the environmental impact of products, but that it has come be considered generally applicable to influencing users' behavior for other purposes (Lockton et al. 2010).

User-Centered Design

User-Centered Design is a design approach in which the users' needs, wants, and limitations to products are given extensive attention at each stage of the design process. As the name suggests, it puts users at center stage during the entire design process, with the aim of improving the quality of the interaction between the user and the product. Instead of focusing on technical features of components, it takes solutions that fit the user as a starting point. It also measures product quality from a user's point of view, taking into account needs, wishes, characteristics, and abilities of the projected user group (Wever et al. 2008). This approach requires designers to anticipate how users are likely to use a product, and to verify the validity of their assumptions by means of tests involving actual users.

The whole design process using this approach includes four major steps: "requirements gathering" (to understand and specify the context of use, including the people who will use the product, what they will use it for, and under what conditions they will use it), "requirements specification" (to specify and identify any user goals that must be met for the product to be successful), "design" (to produce designs and prototypes, building from a rough concept to a complete product), and "evaluation" (to carry out user-based assessment of the site). These four stages are carried out in an iterative fashion, the cycle being repeated until the project's usability objectives have been attained.

There are usually two basic strategies that designers can use in the design process. The first is "functionality matching" design, which means "to adapt a product better to the actual use by consumers and thereby try to minimize negative side effects, i.e. work towards eliminating mismatches between delivered functionalities and desired functionalities" (Wever et al. 2008). In our daily lives, the delivered functionalities of many products are not well matched with our actual desires. For example, often we have multiple appliances switched on but only use part of their functionality. For example, we switch on the TV to hear the news while cooking or doing housework; thus, adding a sound-only mode to televisions might deliver a desired functionality in an eco-efficient way.

The other strategy is "behavior inducing" design, which means to influence the users' behavior actively through product design. There are three methods available for the designers to achieve this goal: eco-feedback, scripting, and forced-functionality (Wever et al. 2008). For example, if we don't switch off the TV fully, but just leave it on stand-by, it will still consume electricity. If the designer could show the message "It is still consuming power" on the screen immediately after the TV is switched to stand-by, I think many people would take notice. This case is the application of eco-feedback. Comparing these two strategies, we find that the latter is more intrusive than the former. Which one should be chosen by the designer depends on the special situation he or she is confronting.

Emotional Design

As was mentioned above, one reason why people dispose of products is that their emotional attachment to them becomes weak. The way to address this is to strengthen such attachment. In a study discussed in their book *The Meaning of Things: Domestic Symbols and the Self,* Mihalyi Csíkszentmihályi and Eugene Rochberg-Halton investigated how things acquired meaning for people and why people loved things. They concluded that meaning was generated by active interaction between people and things. Three factors could contribute the emotional attachments between people and things: the memories that clung to them, the experiences connected to the use of these objects, and references to immediate family (Csíkszentmihályi and Rochberg-Halton 1981).

Emotional Design is an approach to enhance the emotional bond between consumers and products based on these findings. In his 2005 book *Emotional Design,* Don Norman introduced three forms of design to establish emotional relationships between products and their users. The first form is *visceral design*. It refers to users' first impressions of the design outcomes and the emotional responses institutively given. The principles underling visceral design are wired in, consistent across people and cultures: It must be always be attractive. The second form is *behavioral design*. It refers to users' consumption actions based on the emotions raised by the design outcomes. In this level, appearance doesn't really work; only performance matters. Good behavioral design should be human-centered, focusing upon understanding and satisfying the needs of the people who actually use the products. The third form is *reflective design*. It refers to users' reflection on the consumption experiences. If attractiveness comes from the visceral level, then beauty comes from the reflective level. It is influenced by knowledge, learning, and culture. Objects with an unattractive surface can give pleasure; for example, discordant music or ugly art can also be beautiful (Norman 2005).

The practices of the company Eternally Yours give us a good example of emotional design. Instead of the usual emphasis on the technological lifetime and economic lifetime of the products, they focus on the psychological lifetime with the aim to strength consumers' emotional attachments to things. They named this approach "culture durability." They developed three operable strategies to achieve this goal: "Shape 'n Surface," "Sales 'n Services," and "Signs 'n Scripts." The first strategy refers to the forms and materials of the products, such as by seeking materials whose aging process does not render them unattractive. The second refers to the creation of possibilities for improving the support and repair of products, such as cleaning and repair as well as other services and activities that could enhance product longevity. The third refers to the sign-character of products and their scripts, for example, products can feature in stories, which give products more character allowing us to experience them as relevant to ourselves (Verbeek 2005).

Conclusions

Essentially, the both/and approach to ecodesign implies that the work of designers is not just to design green products but also to design sustainable behaviors. In other words, designers now perform a role similar to that of environmental ethicists: persuading humans to behave sustainably. Consequently, the both/and approach loads the designer with ethical responsibilities as their work is no longer a purely technological affair but also a moral matter. Designers provide a material answer to the traditional question of ethics: how to behave (Verbeek 2011). This means that technology can be encoded with morality that can be used as the third way to achieve the goal of regulating people's behaviors by relying on the material force itself. Because of this new constraint mechanism, this third way has some advantages that are not processed by the first two design strategies. For example, the restrictions of morality and law are hysteretic, always lagging behind. If a person violates the law, he or she only can be punished afterward; the constrainsts imposed by materials, on the other hand, are prompt, because technological constraint is always present during its use. Some violations, especially those that have serious adverse effects on society, must be taken under control at the same time when they are carried out. It would be too late if actions are taken against them after damage has been done. In this situation, we can use material constraints to avoid serious illegal behaviors.

When dealing with environmental issues, society tends to focus either on the innovation of green technologies or on promoting sustainable behavior. This either/or approach is problematic. First, if designers ignore the potential influences of the product on the users' behaviors in the design process, it will lead to a rebound effect, which may even offset the environmental merit of the technologies by backfiring. Second, many products are discarded not for their material shortcomings, but for their immaterial shortcomings. When this happens, it does not matter how green a product is. This situation becomes more serious in the era of symbolic consumption, because goods are consumed mainly as symbols. For these reasons, this chapter proposes an integrated both/and approach to ecodesign, in which the relations between users and products takes center stage, instead of focusing only on humans or products.

References

Baudrillard, J. 1998. *The Consumer Society: Myths and Structures*. SAGE.

Bhamra, T. 2004. Ecodesign: The search for new strategies in product development. *Proceedings of the Institution of Mechanical Engineers. Part B, Journal of Engineering Manufacture* 218 (5): 557–569.

Blake, A. 2005. Jevons' paradox. *Ecological Economics* 54 (1): 9–21.

Blake, A. 2008. Historical overview of the Jevons Paradox in the literature. In *The Jevons Paradox and the Myth of Resource Efficiency Improvements*, ed. J. M. Polimeni, K. Mayumi, and M. Giampietro. Earthscan.

Borgmann, A. 1984. *Technology and the Character of Contemporary Life: A Philosophical inquiry*. University of Chicago Press.

Chapman, J. 2009. Design for (emotional) durability. *Design Issues* 25 (4): 29–35.

Csíkszentmihályi, M., and E. Rochberg-Halton. 1981. *The Meaning of Things: Domestic Symbols and the Self*. Cambridge University Press.

Elliott, R. 1994. Exploring the symbolic meaning of brands. *British Journal of Management* 5 (special issue): 13–19.

Jevons, William. 2010. *The Coal Question*. Nabu.

Lockton, D., D. Harrison, and N. Stanton. 2009. Design for sustainable behaviour: Investigating design methods for influencing user behavior. *Design Studies* 30 (6): 704–720.

Lockton, D., D. Harrison, and N. Stanton. 2010. The Design with Intent method: A design tool for influencing user behavior. *Applied Ergonomics* 41 (3): 382–392.

McDonough, D. 2002. *Cradle to Cradle: Remaking the Way We Make Things*. North Point.

Norman, D. 2005. *Emotional Design: Why We Love (or Hate) Everyday Things*. Basic Books.

Van der Ryn, S., and S. Cowan. 1996. *Ecological Design*. Island.

van der Zwan, F., and T. Bhamra. 2003. Alternative function fulfillment: Incorporating environmental considerations into increased design space. *Journal of Cleaner Production* 11 (8): 897–903.

Verbeek, P. P. 2005. *What Things Do—Philosophical Reflections on Technology, Agency, and Design. Penn State*. Penn State University Press.

Verbeek, P. P. 2011. *Moralizing Technology: Understanding and Designing the Morality of Things*. University of Chicago Press.

Wever, R., J. van Kuijk, and C. Boks. 2008. User-centred design for sustainable behaviour. *International Journal of Sustainable Engineering* 1 (1): 9–20.

10 Do We Consume Too Much?

Mark Sagoff

A Roz Chast cartoon in a 2011 issue of *The New Yorker* depicts two robed monks, each carrying a sign. One sign reads "The end of the world is at hand for religious reasons." The other declares "The end of the world is at hand for ecological reasons." Which will it be? Some conservation biologists believe that it might not matter. According to David Orr (2005), there is "an interesting convergence of views between conservation biologists and religious fundamentalists," because "both agree that things are going to hell in the proverbial handbasket." Conservation biologists, Orr notes, often argue that "whether by climate change, biotic impoverishment, catastrophic pollution, resource wars, emergent diseases, or a combination of several, the end is in sight, although we can quibble about the details and the schedule."

Many environmentalists who believe that the world is enjoying its final days subscribe to the Malthusian theory that resources inevitably diminish and become exhausted as population and consumption increase. For many decades, such environmentalists have warned that "human demand is outstripping what nature can supply—even though the great majority of human beings have not even approached the extraordinary American level of resource consumption." They deplore the "human overshoot of the Earth's carrying capacity" (Ehrlich and Ehrlich 2004, 69).

Overconsumption—Ethics, or Economics?

Do we consume too much? To some, the answer is self-evident. If there is only so much food, timber, petroleum, and other material to go around, the more we consume, the less must be available for others. The global economy cannot grow indefinitely on a finite planet. As populations increase and economies expand, natural resources must be depleted; prices will rise, and humanity—especially the poor and future generations at all income levels—will suffer.

Other reasons to suppose we consume too much are less often stated though also widely believed. Of these reasons the simplest—a lesson we learn from our parents and from literature since the Old Testament—may be the best: Although we must satisfy basic needs, a good life is not one devoted to amassing material possessions. What we own comes to own us, keeping us from fulfilling commitments that give meaning to life, such as those to family, friends, and faith. The appreciation of nature also deepens our lives. As we consume more, however, we are more likely to transform the natural world, so that less of it will remain for us to learn from, communicate with, and appreciate.

During the nineteenth century, preservationists forthrightly gave ethical and spiritual reasons for protecting the natural world. John Muir condemned the "temple destroyers, devotees of ravaging commercialism" who "instead of lifting their eyes to the God of the mountains, lift them to the Almighty dollar" (1912, 256). This was not a call for better cost–benefit analysis: Muir described nature not as a commodity but as a companion. Nature is sacred, Muir held, whether or not resources are scarce.

Emerson and Thoreau thought of nature as full of divinity. Walt Whitman celebrated a leaf of grass as no less than the journeywork of the stars: "After you have exhausted what there is in business, politics, conviviality, love, and so on," he wrote in *Specimen Days*, and "found that none of these finally satisfy, or permanently wear—what remains? Nature remains" (1971, 61). These writers thought of nature as a refuge from economic activity, not as a resource for it.

Today many scientists say we are running out of resources or threatening the services ecosystems provide. Predictions of resource scarcity and ecological collapse appear objective and value-free, whereas pronouncements that nature is sacred or has intrinsic value can appear embarrassing in a secular society. One might suppose, moreover, that prudential and economic arguments may succeed better than moral or spiritual ones in swaying public policy. This is especially true if the warnings of resource depletion, global warming, and plummeting standards of living are dire enough—and if a consensus of scientists vouch for them.

Predictions of resource depletion, food scarcity, and falling standards of living, however, may work against our moral intuitions. Consider the responsibility many of us feel to improve the lot of those less fortunate than we. By declaring consumption a zero-sum game, by insisting that what feeds one person is taken from another, environmentalists offer a counsel of despair. Must we abandon the hope that the poor can enjoy better standards of living? The Malthusian proposition that Earth's population already overwhelms its carrying capacity—an idea associated for fifty years with mainstream environmentalist thought—may make us feel guilty but strangely relieves us of responsibility. If there are too many people, some must go.

A different approach, which is consistent with our spiritual commitment to preserve nature and with our moral responsibility to help one another, rejects the apocalyptic narrative of environmentalism. The alternative approach suggests not so much that we consume less as that we invest more. Environmentalists could push for investment in technologies that will increase productivity per unit energy, get more economic output from less material input, provide new sources of power, increase crop yields by engineering better seeds, and move from an industrial economy to a service economy. A really good battery for an electric car could make the petroleum industry nearly obsolete because an electric car could run on a charge that costs pennies per mile. Technological advances of these kinds account for the remarkable improvements in living conditions most people in the world have experienced in the past 40 years, the period over which environmentalists had predicted the steepest declines. They also account for the preservation of nature—for example, the remarkable reforestation of the eastern United States.

What should we environmentalists do? Should we insist, with many conservation biologists and other scientists, that the Earth has reached its limits and the end is in sight, although we can quibble about the details and the schedule? Should we instead leave the End Days to the saints and work with the kinds of knowledge-based high-tech industries that seek to engineer solutions for (or, if necessary, ways to adapt to) the local and global challenges of preserving nature while promoting prosperity?

The idea that increased consumption will inevitably lead to depletion and scarcity, as often as it is repeated, is mistaken both in principle and in fact. It is based on four misconceptions. The first is that we are running out of non-renewable resources, such as minerals. The second is that the world will run out of renewable resources, such as food. The third is that energy resources will soon run out. The fourth misconception argues from the "doubling time" of world population to the conclusion that human bodies cover every inch of the Earth. These misconceptions could turn into self-fulfilling prophecies if we believed them, and if we therefore failed to make the kinds of investments and reforms that have improved standards of living in most of the world.

Are We Running Out of Non-Renewable Resources?

Although commodity markets are volatile (with the markets for petroleum especially sensitive to political conditions), the prices of minerals have declined since the 1980s. The prices of resource-based commodities have declined, and the reserves of most raw materials have increased. The reserves have increased because technologies have greatly improved exploration and extraction (for example, the use of bacteria to leach metals from low-grade ores). Reserves of resources "are actually functions of technology," one analyst has written. "The more advanced the technology, the more reserves become known and recoverable." (Lee 1989, 116) For this reason, among others, as the World Bank reiterated in 2009, although commodity prices are volatile, "over the long run, demand for commodities is not expected to outstrip supply" (World Bank 2009, xi).

One reason for the persistent decline in the costs of minerals and metals is that plentiful resources are quickly substituted for those that become scarce. As technologies that use more abundant resources do the work of technologies dependent on less-abundant resources (for example, ceramics in place of tungsten, fiber optics in place of copper wire, and aluminum cans in place of tin ones), the demand for and the price of scarce resources decline. One can easily find earlier instances of substitution. Early in the nineteenth century, whale oil was the preferred fuel for household illumination. A dwindling supply prompted innovations in the lighting industry, including the invention of gas and kerosene lamps and Edison's carbon-filament electric bulb. Whale oil has substitutes, such as electricity and petroleum-based lubricants. From an economic point of view, technology can easily find substitutes for whale products. From an aesthetic, ethical, and spiritual point of view, in contrast, whales are irreplaceable.

The more we learn about materials, the more efficiently we use them. The progress from whale oil to candles to carbon-filament incandescent lamps to tungsten incandescent lamps, for example, decreased the energy required for and the cost of a unit of household lighting by many times. On perfecting the electric bulb, which made lighting inexpensive, Thomas Edison is widely quoted as saying that "only the rich will burn candles." Compact fluorescent lights are four times as efficient as today's incandescent bulbs and last ten to twenty times as long. Comparable energy savings are available in other appliances; for example, refrigerators sold today are more efficient than those sold in 1990, saving consumers billions of dollars on their electric bills. If the future is like the past, the productivity of natural resources will continue to rise along with the productivity of labor, and we will require fewer resources per unit of production.

Modern economies depend more on the progress of technology than on the exploitation of nature. Although raw materials will always be necessary, knowledge has become the essential factor in the production of goods and services. Technological advance, which seems to be exponential insofar as each discovery prompts others, promises to improve standards of living while lightening the human footprint on the natural world. Of course, no one believes that economic development (or technological and scientific progress) will automatically lead to environmental improvement. It only provides the means; we must gather the moral, cultural, and political will to pursue the end. We can always obtain other resources. The limits to knowledge are the limits to growth.

Will There Be Enough Food?

"People today," a prominent agricultural economist wrote in 2000, "have more adequate nutrition than ever before and acquire that nutrition at the lowest cost in all human history, while the world has more people than ever before—not by a little but by a lot." (Johnson 2000, 1) This happened, Johnson argued, because "we have found low-cost and abundant substitutes for natural resources important in the production process" (ibid., 2). By around 2000, the price of food and feed grains, in real dollars (adjusted for inflation), had declined by half from what it had been 50 years earlier in international markets.

From 1961 to 2009, global production of food doubled (FAO 2009). The world produces enough cereals and oilseeds to feed a healthful vegetarian diet adequate in protein and calories to 10 billion people—a billion more than the number at which demographers predict world population will peak later this century. If, however, the idea is to feed 10 billion people not healthful vegetarian diets but the kind of meat-laden, artery-clogging, obesity-causing gluttonous meals that many Americans eat, the production of grains and oilseeds may have to triple—primarily to feed livestock (Matson and Vitousek 2006, 709). Conceivably, if everyone had the money to pay for food at current prices, with technological advances occurring particularly in bioengineering, the world could produce enough beef and donuts to fatten everyone for the slaughter of diabetes, cirrhosis, and heart disease.

Farmers worldwide could double the acreage in production, but this should not be necessary. Increasing productivity will flow from the agricultural revolution driven by biotechnology—a field that includes advanced genetics and genomics, bioinformatics, genetically modified plants, and tissue culture. According to Lester Brown (1989), "there are vast opportunities for increasing water efficiency" in arid regions, ranging from installing better water-delivery systems to planting drought-resistant crops," and "scientists can help push back the physical frontiers of cropping by developing varieties that are more drought resistant, salt tolerant, and early maturing. The payoff on the first two could be particularly high." Biotechnology introduces an entirely new stage in humankind's attempts to produce more crops and plants. The Gene Revolution takes over where the Green Revolution left off.

Before one heads to the nearest steak house to tuck into a T-bone, one should acknowledge three problems for this optimistic account. First, the essential input onto agriculture is money. Money is not spread evenly over the Earth; it is concentrated in the wealthier nations. According to the Millennium Ecosystem Assessment (2005), "despite rising food production and falling food prices, more than 850 million people still suffer today from chronic undernourishment." Many of the poorest countries, such as Chad and Congo, possess more than enough excellent agricultural land but lack social organization and investment. Institutional reform—responsible government, peace, the functioning of markets, the provision of educational and health services (in other words, development)—is the appropriate response to poverty and therefore the appropriate response to malnutrition. Second, according to the Millennium Ecosystem Assessment, "among industrial countries, and increasingly among developing ones, diet-related risks, mainly associated with overnutrition, in combination with physical inactivity now account for one third of the burden of disease"; by comparison, "worldwide, undernutrition accounted for nearly 10% of the global burden of disease." Third, to make 9 billion people obese, biotech-based agriculture would have to convert the Earth to a feedlot for human beings. Farmers can now provide a healthful diet for that many people on less acreage than they use today, thus sparing land for nature. In other words, we can spare nature by sparing ourselves.

By locking themselves into the Malthusian rhetoric by predicting impending worldwide starvation and using the plight of the very poor as evidence of it, environmentalists ignore and even alienate groups who emphasize the quality and safety rather than the abundance of food and who understand that under-nutrition represents a local not a global problem. The discussion has moved from the question whether the Earth sets limits to the question of how to get wealthy people to eat less and poor people to eat more. Advocates of animal rights deplore horrific the feedlot operations and the related factory-farm methods required to overfeed people. Environmentalists have obvious allies in advocates of human development, public health, and animal rights. To have any credibility, however, environmentalists must abandon the apocalyptic narrative.

Are We Running Out of Energy?

Predictions that the world would by now have run out of petroleum, or will do so shortly, are an industry. Among the titles of books published in the early 2000s were *Beyond Oil: The View from Hubbert's Peak, The End of Oil: On the Edge of a Perilous New World, Out of Gas: The End of the Age of Oil,* and *The Party's Over: Oil, War and the Fate of Industrial Societies* (Deffeyes 2005; Roberts 2005; Goodstein 2004; Heinberg 2003). The most persistent worries about resource scarcity concern energy. "The supply of fuels and other natural resources is becoming the limiting factor constraining the rate of economic growth," a group of experts proclaimed in 1996 (Gever et al. 1996, 9). They predicted the exhaustion of domestic oil and gas supplies by 2020 and, within a few decades, "major energy shortages as well as food shortages in the world."

In stark contrast with the dire jeremiads of the 1990s, the US Department of Energy 2012 projections "show natural gas and renewables gaining an increasing share of US electric power generation, domestic crude oil and natural gas production growing, reliance on imported oil decreasing, US natural gas production exceeding consumption, and energy-related carbon dioxide emissions remaining below their 2005 level through 2035" (US Energy Information Administration 2012). According to the Department of Energy, oil production had not peaked in the US, and that "domestic crude oil production is expected to grow by more than 20 percent over the coming decade" (ibid.). They projected that increased oil, natural gas and renewable energy production and energy efficiency improvements would significantly reduce the United States' reliance on imported energy.

The most abundant fossil or carbon-based fuel is coal, and some of the largest reserves of it are found in the United States. They will last more than 100 years. In this respect, no global shortages of hydrocarbon fuels are in sight. There is no immediate danger of the entire world's running out of energy. That is not what the energy problem is all about. Yet for decades environmental Cassandras have reiterated that we are running out of energy, thus directing attention to sources rather than sinks. Thankfully there is a growing consensus that the real energy problems are global climatic instability and global political instability. Reasonable minds can disagree about which problem is worse; but both require that the world move away from its dependence on fossil fuels and toward reliance on cleaner and smarter kinds of energy and toward more efficient use of energy.

First, the burning of hydrocarbon fuels contributes to global warming and climate change. In 1958, the concentration of carbon dioxide (CO_2) stood at 315 parts per million (ppm). Today, it has reached 394 ppm, about one third higher than the historical norm over 400,000 years. Levels of CO_2 are increasing so fast that in 40 or 50 years concentrations may be twice the historic levels (www.climate.nasa.gov/climate_resources). Since the planetary climate may already be changing in response to current CO_2 loadings, scientists consider the situation urgent. The global energy problem has less to do with depleting resources than with controlling emissions.

The second problem has to do with geopolitical stability. Thomas Friedman (2006) observes that oil-rich states tend to be the least democratic, and that the wealthier the ruling class gets, the more tyrannical, truculent, obstructive, and dangerous it becomes. The "petrocracies" destabilize global balances of power while holding oil-dependent states hostage. Although the food problem is best understood as local (giving the very poor access to nutrition), the energy problem is global. The principal concern is not the supply of energy but the effects of its use on geopolitics and climate.

Although leading environmentalists have focused on scarcity, they have also joined nearly everyone else in deploring the effects of the consumption of carbon-based fuels on the political and the atmospheric climate. To provide leadership and direction rather than simply reiterate their apocalyptic projections, environmentalists should advocate investment in some mix of power-producing and climate-sparing technologies. There is a smorgasbord of suggestions. These include hybrid, plug-in hybrid, and electric vehicles; greater energy efficiency in housing and appliances; and the production of liquid fuels from renewable sources, some produced by genetically engineered or synthesized microorganisms capable of creating biomass cheaply or even directly splitting the carbon dioxide molecule. Other approaches include the expansion of nuclear power generation (including smaller distributed and sealed units), the development of geothermal and wind energy, and basic and applied research in battery technology, fuel-cell technology, tidal power, and other forms of power. Efforts are underway to construct a smarter and more efficient electric energy transmission grid.

In the American Reinvestment and Recovery Act of 2009, the Obama administration threw a staggering amount of money at clean energy technologies. It is impossible at this time to pick winners among the scores of innovations. Some of this money will stick. Commercially available technologies can support present or greatly expanded worldwide economic activity while stabilizing global climate—and can save money. Even very large expansions in population and industrial activity need not be energy-constrained.

If many opportunities exist for saving energy and curtailing pollution, why have we not seized them? One reason is that low fossil-fuel prices remove incentives for fuel efficiency and for moving to other energy sources. If energy supplies were scarce, prices would have risen to levels that would force the kinds of innovations and transitions we now need the political will to make. Environmentalists might have more credibility in supporting novel forms of energy production if they were not weighed down by decades of doomsaying. The major obstacles standing in the way of a clean-energy economy are not technical in nature but concern the regulations, incentives, public attitudes, and other factors that make up the energy market.

Are There Too Many People?

In the 1970s, the population crisis was easy to define and dramatize. The Malthusian logic of exponential growth or "doubling times," forcefully presented in books such as *The Population Bomb* and *The Population Explosion*, argued that the "battle to feed all of humanity is over" and analogized the spread of population with cancer: "A cancer is an uncontrolled multiplication of cells; the population explosion is an uncontrolled multiplication of people. ... The [surgical] operation will demand many apparently brutal and heartless decisions. The pain may be intense. But the disease is so far advanced that only with radical surgery does the patient have a chance of survival." (Ehrlich 1971, 152)

By emphasizing the exponential mathematics of population growth—as if people were cancerous cells whose reproductive freedom had to be controlled by radical surgery—environmentalists made four mistakes.

First, they missed the opportunity to endorse the belief that people should have all—but only—the children they want. The goal of assisting parents worldwide to plan for their children might appeal to family values and thus to social conservatives in a way that concerns about too many people did not. Efforts to improve the status of women may enjoy more political support and may be more effective than conventional fertility-control policies.

Second, by inveighing against economic growth (by demanding a small economy for a small Earth), environmentalists alienated potential allies in the development community. Leading environmentalists explicitly rejected the hope that development can greatly increase the size of the economic pie and pull many more people out of poverty. This hope, Paul and Anne Ehrlich wrote, expresses a "basically a humane idea ... made insane by the constraints nature places on human activity" (1990, 269).

Development economists replied that a no-economic-growth approach in the developing world would deprive entire populations of access to better living conditions and lead to even more deforestation and land degradation. Amartya Sen, among other scholars, pointed out that insistence on the Malthusian belief that nature puts narrow constraints on human activity diverts attention from the real causes of malnutrition, namely poverty and political powerlessness. The Malthusian approach, Sen argued, leads to complacent optimism because food production at the global level is more than adequate. With such "misleading variables as food output per unit of population, the Malthusian approach profoundly misspecifies the problems facing the poor of the world," which have to do with local conditions not with global constraints, and "it is often overlooked that what may be called 'Malthusian optimism' has actually killed millions of people" (Sen 1989).

Third, by invoking "doubling times" as if that concept could be as meaningfully applied to people as to tumors, environmentalists ignored science and reason—that is, everything demographers knew about the transition then underway to a stable global population. As people move to cities, where children are not needed to do agricultural labor, as they are assured that their children will survive (so they can have fewer children), and as the status of women improves, families become smaller. World population growth, which resulted from lower mortality not higher fertility, had been decelerating since the 1950s and dramatically after the 1970s. In 2008, the United Nations projected the global population "to reach 7 billion in late 2011, up from the current 6.8 billion, and surpass 9 billion people by 2050," when it would stabilize and probably decline (UN Department of Economic and Social Affairs, Population Division, 2009, vii). Most demographers believe that population will stop increasing during this century and then decline slowly to perhaps 8.4 billion in 2100 (ibid., vii).

Today, most people live in countries or regions in which fertility is below the level of long-run replacement. According to a United Nations press release published in 2011, "42 percent of the world's population lives in low-fertility countries, that is, countries where women are not having enough children to ensure that, on average, each woman is replaced by a daughter who survives to the age of procreation." Another 40 percent live in nations approaching that level (UN Press Release 2011). Of course, population will increase inexorably as lifespans increase. According to UN projections, "globally, life expectancy is projected to increase from 68 years in 2005-2010 to 81 in 2095–2100" (ibid.).

Fourth, the environmental community has yet to respond to the principal moral problem that confronts population policy—one that involves longevity not fertility. The oldest segments of the population increase the fastest as science and technology extend the length of life. A UN report observes that in developed regions of the world "the population aged 60 or over is increasing at the fastest pace ever (growing at 2.0 percent annually) and is expected to increase by more than 50 per cent over the next four decades, rising from 264 million in 2009 to 416 million in 2050," and that developing world is aging even more rapidly: "Over the next two decades, the population aged 60 or over in the developing world is projected to increase at rates far surpassing 3 per cent per year, and its numbers are expected to rise from 473 million in 2009 to 1.6 billion in 2050" (ibid., viii).

Anyone interested in doubling times or exponential growth should consider the following statistics: In industrialized countries the number of centenarians has doubled every decade since 1950. In many countries, people aged 80 or over constitute the fastest-growing segment of the population. In 1900, 374,000 people in the United States had attained the age of 80;. Today, 10 million Americans are elderly; by 2030, that number is expected nearly to double, making huge demands on younger workers, whose labor may be needed and whose incomes will be taxed to pay for their care.

The problem is no longer Malthus, it's Methuselah. What do environmentalists say about this? As long as environmental leaders argue forever that there are too many people without suggesting how long a life should last, they seem self-serving. These environmentalists appear to comprise a vast and growing gerontocracy outraged that younger people whom they may need to take care of them presume to care for their own children.

What Is Wrong with Consumption?

Many of us who grew up with the attitudes of the 1960s and the 1970s took pride in how little we owned. We celebrated our freedom when we could fit all our possessions into the back of a car. As we grow older, we tend to accumulate an appalling amount of stuff. Piled high with gas grills, lawn mowers, excess furniture, bicycles, children's toys, garden implements, ladders, lawn and leaf bags stuffed with memorabilia, and boxes yet to be unpacked from the last move, the two-car garages beside our suburban homes are too full to accommodate our SUVs. The quantity of resources (particularly energy) we waste and the quantity of trash we throw away add to our worries.

We are distressed by the suffering of others, the erosion of the ties of community, and the loss of the beauty and spontaneity of the natural world. These concerns express the most traditional and fundamental of American religious and cultural values. Even if predictions of resource depletion and ecological collapse are mistaken, it seems that they ought to be true, to punish us for our success and our sins.

Perhaps a feeling of guilt impels environmentalists to adopt their vision of impending Apocalypse, in the form of imminent resource depletion, starvation, and ecological Armageddon. In contrast, religious communities, especially mainstream Evangelical and other Christian groups, emphasize stewardship of the Earth for the very long run. In fact, more than sixty faith-based groups today pursue missions they describe as "environmental conservation" or "caring for creation" (www.webofcreation.org). In 1990, the National Association of Evangelicals issued a policy document that urged greater concern for the environment that included this statement:

> We urge Christians to shape their personal lives in creation-friendly ways: practicing effective recycling, conserving resources, and experiencing the joy of contact with nature. We urge government to encourage fuel efficiency, reduce pollution, encourage sustainable use of natural resources, and provide for the proper care of wildlife and their natural habitats. (National Association of Evangelicals 1990)

If the environmental community were to join with mainstream religious groups in preaching a narrative of hope rather than one of futility and imminent demise, the environmental movement would find itself in a better position to work with charitable organizations to relieve the lot of the poorest of the poor. There is a lot of misery worldwide to relieve.

However, imposing a market economy on traditional cultures in the name of development—the idea that everyone can and should always produce and consume more—is not always the solution. It creates problems as well as opportunities. A market economy may dissolve the ties to family, land, community, and place on which indigenous peoples traditionally rely for their security. Thus, projects intended to relieve the poverty of indigenous peoples may, by causing the loss of cultural identity, engender the very powerlessness they aim to remedy. Pope Paul VI, in the encyclical Populorum Progressio (1967), described the dilemma confronting indigenous peoples: "either to preserve traditional beliefs and structures and reject social progress; or to embrace foreign technology and foreign culture, and reject ancestral traditions with their wealth of humanism." A similar dilemma confronts wealthy societies. No one has written a better critique of the assault that commerce makes on the quality of our lives than Thoreau provides in *Walden*. We are always in a rush—a "Saint Vitus' dance" (Thoreau 1995, 174). Idleness is suspect. Americans today spend less time with their families, neighbors, and friends than they did in the 1950s. We are almost literally working ourselves to death. That money does not make us happier, once our basic needs are met, is a commonplace overwhelmingly confirmed by sociological evidence. Paul Wachtel, who teaches social psychology at the City University of New York, has concluded that bigger incomes "do not yield an increase in feelings of satisfaction or well-being, at least for populations who are above a poverty or subsistence level" (Wachtel 1994, 5). This cannot be explained simply by the fact that people have to work harder to earn more money. Even those who win large sums of money in lotteries often report that their lives are not substantially happier as a result (Argyle 1986). Well-being depends upon health, membership in a community in which one feels secure, friends, faith, family, love, and virtues that money cannot buy.

Economists in earlier times predicted that wealth would not matter to people once they attained a comfortable standard of living. "In ease of body and peace of mind, all the different ranks of life are nearly upon a level," wrote Adam Smith, the eighteenth-century English advocate of the free market (1976, 185). In the 1930s the British economist John Maynard Keynes argued that after a period of expansion accumulation of wealth would no longer improve personal well being (Keynes 1963). Subsequent economists, however, found that, even after much of the industrial world had attained the levels of wealth Keynes thought were sufficient, people still wanted more. From this they inferred that wants are insatiable (Nelson 1991). Perhaps this is true. But the insatiability of wants poses a difficulty for standard economic theory, which posits that humanity's single goal is to increase or maximize wealth. If wants increase as fast as income grows, what purpose can wealth serve?

Whether or not economic growth is sustainable, there is little reason to think that continued growth is desirable once people attain a decent standard of living. It is no longer possible for most people to believe that economic progress will solve all the problems of mankind, spiritual as well as material. Environmentalists will not make convincing arguments as long as they frame the debate over sustainability in terms of the physical limits to growth rather than the moral purpose of it. Even if technology overcomes the physical limits nature sets on the amount we can produce and consume, however, there are moral, spiritual, and cultural limits to growth. Environmentalists defeat themselves by denying the power of technological progress. If the debate were couched not in economic terms but in moral or social terms—if it were to center on the values we seek to serve rather than the resources we may exhaust—environmentalists might more easily win the argument.

Making a Place for Nature

According to Thoreau, "a man's relation to Nature must come very near to a personal one" (1949, 252). For environmentalists in the tradition of Thoreau and Muir, stewardship is a form of fellowship. Although we must use nature, we do not value it primarily for the economic purposes it serves. We take our bearings from the natural world—our sense of time from its days and seasons, and our sense of place from the character of a landscape and the particular plants and animals native to it. An intimacy with nature ends our isolation in the world. We know where we belong, and we can find the way home.

In defending old-growth forests, wetlands, or species, we environmentalists make our best arguments when we think of nature chiefly in aesthetic and moral terms. Rather than having the courage of our moral and cultural convictions, however, we too often rely on economic arguments for protecting nature, in the process attributing to natural objects more instrumental value than they have. By imputing to an endangered species an economic value or a price much greater than it fetches in a market, we "save the phenomena" for economic theory but do little for the environment. When we make the prices come out "right" by imputing market demand to aspects of nature, which in fact have moral, spiritual, or aesthetic value, we confuse ourselves and fail to convince others.

There is no credible argument that all or even most of the species we are concerned to protect have any economic significance or that they are essential to the functioning of the ecological systems on which we depend. If whales were to become extinct, for example, the seas would not fill up with krill. David Ehrenfeld, a biologist at Rutgers University, points out that the species most likely to be endangered are those the biosphere is least likely to miss: "Many of these species were never common or ecologically influential; by no stretch of the imagination can we make them out to be vital cogs in the ecological machine." (Ehrenfeld 1988, 215)

Species may be profoundly important for cultural and spiritual reasons, however. Consider the example of the wild salmon, whose habitat is being destroyed by hydroelectric dams along the Columbia River. Although this loss is not important to the economy overall (there is no shortage of farmed salmon), it is of great cultural significance to the Amerindian tribes that have traditionally subsisted on wild salmon, and to the region as a whole. By viewing local flora and fauna as a sacred heritage—by recognizing their intrinsic value—we discover who we are rather than what we want. On moral and cultural grounds society might be justified in making economic sacrifices—removing dams, for example—to protect remnant populations of the Snake River sockeye, even if, as critics complain, hundreds or thousands of dollars are spent for every fish.

Even those plants and animals that do not define places possess enormous intrinsic value and are worth preserving for their own sake. What gives these creatures value lies in their histories, wonderful in themselves, rather than in any use to which they can be put. The biologist E. O. Wilson elegantly takes up this theme: "Every kind of organism has reached this moment in time by threading one needle after another, throwing up brilliant artifices to survive and reproduce against nearly impossible odds." (1992, 345. Every plant or animal evokes not just sympathy but also reverence and wonder in those who know its place, properties, and history.

In *Earth in the Balance*, Al Gore wrote "We have become so successful at controlling nature that we have lost our connection to it" (1992, 225). It is all too easy, Gore wrote, "to regard the earth as a collection of 'resources,' having an intrinsic value no larger than their usefulness at the moment" (ibid., 1). The question before us is not whether we are going to run out of resources. It is whether the theory of welfare economics is the appropriate context for thinking about environmental policy.

Even John Stuart Mill, one of the principal authors of utilitarian philosophy, recognized that the natural world has great intrinsic and not just instrumental value. More than 100 years ago, as England lost its last wild places, Mill condemned a world "with nothing left to the spontaneous activity of nature; with every rood of land brought into cultivation, which is capable of growing food for human beings; every flowery waste or natural pasture ploughed up; all quadrupeds or birds which are not domesticated for man's use exterminated as his rivals for food, every hedgerow or superfluous tree rooted out, and scarcely a place left where a wild shrub or flower could grow without being eradicated as a weed in the name of improved agriculture" (1987, 750).

The world has the wealth and the resources to provide everyone the opportunity for a decent life. We consume too much when market relationships displace the bonds of community, compassion, culture, and place. We consume too much when consumption becomes an end in itself and makes us lose affection and reverence for the natural world.

References

Argyle, Michael. 1986. *The Psychology of Happiness*. Methuen.

Brown, Lester. 1989. The grain drain. *Futurist* 23 (4): 17–18.

Deffeyes, Kenneth S. 2005. *Beyond Oil: The View from Hubbert's Peak*. Farrar, Straus and Giroux.

Ehrenfeld, David. 1988. Why put a value on biodiversity? In *Biodiversity*, ed. E. O. Wilson. National Academy Press.

Ehrlich, Paul. 1971. *The Population Bomb*. Ballantine Books.

Ehrlich, Paul R., and Anne H. Ehrlich. 1990. *The Population Explosion*. Simon and Schuster.

Ehrlich, Paul R., and Anne H. Ehrlich. 2004. *One with Nineveh: Politics, Consumption, and the Human Future*. Island.

FAO (UN Food and Agriculture Organization). 2009. Global food supply gradually steadying. Press release, June 4 (http://www.fao.org/news/story/en/item/20351/icode/).

Friedman, Thomas L. 2006. "As energy prices rise, it's all downhill for democracy." *New York Times*, May 5.

Goodstein, David. 2004. *Out of Gas: The End of the Age of Oil*. Norton.

Gever, John. Robert Kaufmann, David Skole, and Charles Vorosmarty. 1996. *Beyond Oil: The Threat to Food and Fuel in the Coming Decades*. Ballinger.

Gore, Al. 1992. *Earth in the Balance: Ecology and the Human Spirit*. Houghton Mifflin.

Heinberg, Richard. 2003. *The Party's Over: Oil, War and the Fate of Industrial Societies*. New Society Publishers.

Johnson, Gale D. 2000. Population, food, and knowledge. *American Economic Review* 90 (1): 1–14.

Keynes, John Maynard. 1963. Economic possibilities for our grandchildren. In Keynes, *Essays in Persuasion*. Norton.

Lee, Thomas H. 1989. Advanced fossil fuel systems and beyond. In *Technology and Environment*, ed. Jesse H. Ausubel and Hedy E. Sladovich. National Academy Press.

Matson, Pamela, and Peter Vitousek. 2006. Agricultural intensification: Will land spared from farming be land spared for nature? *Conservation Biology* 20 (3): 709–710.

Mill, John Stuart. 1987. *Principles of Political Economy with some of their Applications to Social Philosophy*. Kelley.

Assessment, Millennium Ecosystem. Ecosystems and Human Well-being: Current State and Trends. 2005. http://www.millenniumassessments.org/.

Muir, John. 1912. *The Yosemite*. Century.

National Association of Evangelicals. 1990. *Stewardship*. nae.net

Nelson, Robert H. 1991. *Reaching for Heaven on Earth: The Theological Meaning of Economics*. Rowman and Littlefield.

Orr, W. David. 2005. Armageddon versus extinction. *Conservation Biology* 19 (2): 290–292.

Pope Paul VI. 1967. "Populorum Progressio," Encyclical of Pope Paul VI on the Development of Peoples, March 26, 1967. In *The Papal Encyclicals 1958–1981*, ed. Claudia C. Ihm. Pierian Press.

Roberts, Paul. 2005. *The End of Oil: On the Edge of a Perilous New World*. Houghton Mifflin.

Sen, Amartya. 1989. *Resources, Values, and Development*. Harvard University Press.

Smith, Adam. 1976. *The Theory of the Moral Sentiments*, ed. D. D. Raphael and A. L. Macfie. Clarendon.

Thoreau, Henry David. 1949. *The Journal of Henry David Thoreau*. vol. 10. Ed. B. Torrey and F. Allen. Houghton Mifflin.

Thoreau, Henry David. 1995. *Walden: Or, Life in the Woods*. Dover.

UN Department of Economic and Social Affairs, Population Division. 2009. World Population Prospects: The 2008 Revision, Highlights, Working paper.

UN Press Release. May 3, 2011. World Population to reach 10 billion by 2100 if Fertility in all Countries Converges to Replacement Level.

US Energy Information Administration. 2012. Annual Energy Outlook 2012 with Projections to 2035. www.eia.gov/forecasts/aeo

Wachtel, Paul. 1994. Consumption, satisfaction, and self-deception. Paper presented at conference on Consumption, Stewardship and the Good Life, University of Maryland.

Whitman, Walt. 1971. *Specimen Days*. Godine.

Wilson, Edward O. 1992. *The Diversity of Life*. Harvard University Press.

World Bank. 2009. Global Economic Prospects. www.climate.nasa.gov/climate_resources www.webofcreation.org

11 Sustainable Technologies for Sustainable Lifestyles

Philip Brey

An adequate approach to technology is unquestionably a major component of any strategy toward sustainable development. The widespread production and use of modern technology is a defining feature of industrial society. Without modern technology, there probably would not be a problem of sustainability to begin with. Many sources of pollution and environmental degradation are results of large-scale development and use of modern technology, including the extraction, processing, and consumption of fossil fuels, the large-scale dissemination of chemical pollutants, the production of non-biodegradable waste such as plastics, glass, and pesticides, and soil degradation caused by modern mechanized agriculture. At the same time, technology is also a major factor in any solution to environmental problem. Any such solution will have to consider how technologies can be made more ecological and sustainable, and how new technologies can be developed to mitigate environmental pollution and degradation. This chapter aims to investigate the importance of sustainable technology as part of a strategy for moving toward sustainable development, and to analyze how sustainable technology would look.

There is now broad agreement among the nations of the world that economic development should be environmentally sustainable. It is also recognized that industrial societies are the main contributors to environmental degradation. Since the 1980s, the explicit, shared goal of many countries around the world for environmental policy has been *sustainable development*. That goal has guided most international agreements on the environment and climate change, including the Kyoto Protocol and the Copenhagen Accord, and is guiding national strategies and the approaches of industry and environmental organizations.

Sustainable development was defined in the influential 1987 report of the World Commission on Environment and Development (the Brundtland report) as "development that meets the needs of the present generation without compromising the ability of future generations to meet their own needs" (WCED 1987, 43). This definition can be read as expressing basic values that should underlie economic development policies. Taking seriously the needs of future generations, if not the integrity of nature as a whole, requires at least that "the environment should be protected in such a condition and to such a degree that environmental capacities (the ability of the environment to perform its various functions) are maintained over time" (Jacobs 1991, 79). This requirement, it is generally agreed, implies serious reductions in the generation of substances and gases that threaten ecosystemic life cycles and in the immediate destruction or consumption of elements of nature, and serious efforts to protect ecosystems and natural resources.

Although there is agreement on sustainable development as a goal, there is less agreement on the right strategy for attaining it. Should the emission of pollutants be regulated through a cap-and-trade scheme, or should alternative reduction solutions be sought? Should renewable energy be heavily subsidized, or should its development be left to the market? Should we start phasing out the production and use of coal and oil now, or should we continue to use them? Yet, as I have argued before (Brey 1997), despite all this disagreement there are many shared assumptions. Most Western nations agree on a basic strategy that has been called *ecological modernization*. Ecological modernization is an environmental control strategy that aims at "greening" production processes and the global economy in a way that leaves existing institutions and practices intact as much as possible. It centrally involves a transformation of technology: Industrial production processes and produced artifacts are to become environmentally efficient or environmentally friendly, and the production and consumption of technological artifacts should use up fewer nonrenewable resources, emit fewer greenhouse gases, and produce less harmful waste, both in production and consumption cycles.

In this chapter, I will present the strategy of ecological modernization, and I will subject it to a critique, arguing that its reform of technology, the economy system, and production are consumption are likely to be insufficient for sustainable development. The main problem with the strategy, I will argue, is that it only aims at modest reforms in the institutions of modern industrial society, without any fundamental reform. The fundamental reform that is needed is in our values regarding the quality of life and how to achieve it. Modern Western lifestyles are based on ideals of consumption that are unsustainable unless they are thoroughly reformed. I will therefore present an alternative second strategy for sustainable development, and correspondingly for sustainable technology, that focuses on the cultural-environmental reform of consumption and of the lifestyles and conceptions of the quality of life presupposed by it. I will discuss how this strategy relates to ecological modernization and to existing attempts to reform consumption and lifestyles, including the voluntary simplicity and degrowth movements. I will also discuss how this approach may transform the development and the use of technology.

Ecological Modernization as a Path toward Sustainability

Many nations now have national strategies for sustainable development that typically center around reduction in greenhouse gas emissions through conservation, efficient use of raw materials, waste reduction, the use of biodegradable waste materials, better use of land, protection of biodiversity, and a shift toward renewable energies. There have also been efforts to develop such strategies at an international level, for example in the Kyoto Protocol and the Copenhagen Accord. What these strategies have in common is that they aim to reform production and consumption processes so as to achieve sustainability while also holding on to the basic institutions of modern society, including industrialism (the economic organization of society based on large-scale industries) and a market-based, capitalist economic model that include the ideals of economic growth, limited government intervention, free trade, and consumerism.

This approach finds its academic expression in the theory of *ecological modernization*, which has been developed since the 1980s by Arthur Mol, Gert Spaargaren, Joseph Huber, and others (Huber 1982; Mol 1995; Spaargaren and Mol 1992; Mol, Sonnenfeld, and Spaargaren 2009). Ecological modernization theory explicitly rejects the assumption that a fundamental reorganization of the core institutions of modern society is necessary for sustainable development (Mol and Spaargaren 2000, 19). Instead, it holds that sustainable development can be attained through modest reform of some of these institutions, most notably by incorporation of ecological principles into industry and the economic system. The key to this reform is the development and introduction of new ecological technologies for industrial production.

Ecological modernization (EM) aims to transform industry through a series of measures that control the source of environmental problems, or source-oriented measures. This may involve controlling emissions (by adding technologies that reduce emissions and waste streams without changing responsible processes of production of consumption themselves), volume control (legal and organizational measures that reduce the quantity of base materials and products without limiting the processes of production and consumption), and structural changes (usually of a technological nature) that modify the processes of production and consumption.

Structure-oriented measures tend to come in three kinds. The first is *integrated chain management*. This is a "cradle-to-grave" approach to production that aims to minimize the environmental load of product chains, from extraction to production, by looking at ways of limiting environmental load of phases in the chain without increasing it for other phases. Another is *energy expansion*: the more efficient use of energy in production processes and in products and the use of renewable energy sources. A third is *quality improvement*: the production of more durable goods that can moreover be recycled. More and more, such measures are conditioned by economic measures that stimulate environmental solutions and penalize unsustainable practices.

The explicit embrace by EM of the institutions of modern society is a response to environmental critiques in the 1970s, which argued for a more radical reform of these institutions. Such critiques tended to be highly critical of industrialism, capitalism, and modernity and their ideals of progress and growth, and often advocated a partial or complete dismantling of industrial society and a return to small-scale technologies (Schumacher 1973; Illich 1973). Proponents of EM held that approaches that embraced modernity and industrialism rather than rejecting them were more likely to lead to changes in industrial society that would yield sustainability.

By accepting and upholding the institutions of modern society, EM can be seen as accepting the project of modernization that began in the Industrial Revolution. This process of modernization is characterized by an ideal of progress that is exercised through increases in productivity and technological complexity, rationalization of production, the employment of scientific principle and method, and professionalization within the economic context of free-market capitalism.

The philosophical background of the project of modernization is found in the principles of modernity. The central principle of modernity is the principle of autonomy: the idea that individuals and societies can attain self-determination or self-rule, and can define their own laws of action independently from their environment. Reason and its most successful manifestation, science, were to guarantee this autonomy through the laws and principles they bring forth. The ideal of progress, as another central principle of modernity, is the belief that the employment of reason and its special forms can lead to increases in autonomy and improvements in the human condition. The project of modernization can thus be understood as a project aimed at increasing the autonomy of its agents by granting them increased control over their own destiny through technology and by giving them extended powers to realize their goals and satisfy their desires.

Ecological modernization is the logical answer to the ecological crisis from within the modernization project. It is a control strategy that is starting to replace the more conservative control strategy of end-of-pipe measures, which has turned out to be insufficiently effective in the face of mounting global environmental problems. The new strategy aims at an ecological transformation of the modernization process—that is, a transformation based on ecological principles as developed within the science of ecology. The prime targets of ecological modernization are the institutions of technology and the economy. The technological and economic system is to be made part of the ecological system, and hence to incorporate ecological principles in its own operations. Integral chain management, in which industries strive to imitate life-cycle processes found in nature so as to be ecologically sound, is an example of such a process.

The ecologization of technology is, as I have said, to be attained through structural reform of (agro-)industrial production processes. New technologies (including micro-electronics, genetic engineering, and nanotechnology) and new materials are thought to be able to play a central role in this reform process, because they limit resource inputs, resource use, and emissions (Simonis 1989). The ecologization of the economy (correlating with an economization of ecology) is thought to involve the reform of economic theory and economic policies. Most importantly, a value must be placed on nature, as a force of production, to allow its conservation and protection to be an integral part of economic development strategies. But it may also involve "more incidental eco-taxes, the introduction of environmental liability, the redirection of insurance condition toward environmental care, the increasing demand for ecologically sound products on the market, the introduction of the environment as a factor in economic competition and of environmental audits as a precondition for commercial loans and economic investments" (Mol 1995, 40).

Ecological modernization should be understood as a control strategy defined *within* the general project of modernization because it assumes that the environmental conflict is not inherent to the project of modernization but can be controlled from within it. It leaves the basic tenets of the project of modernization intact, together with the basic institutions and ideals of modernity. This is evident in several ways. Ecological modernization is targeted at a reform of only two institutions of modernity: technology and the economy. Moreover, in spite of the drastic reform of these two institutions implied by ecological modernization, their central principles remain intact. In the ecological reform of economics, the ideal of growth, as an index of progress, is preserved, as is, in most cases, the adherence to free-market capitalism. In the reform of the institution of technology, the aim is not a reduction of the role and influence of technology, or deindustrialization, but rather an increase in the environmental efficiency of technology. The modernist idea that technology should play a central role in solving major problems is, moreover, retained: The control strategy of ecological modernization grants a central role to new technologies in solving environmental problems.

In fact, the project of ecological modernization can largely be understood as a technological control strategy. This can be seen in the fact that a central part of the strategy lies in the technological reform of production systems. But even concomitant changes in the organization of industry and in economic theory and policy can be understood as technological changes when the notion of technology is taken in a broad sense, as the implementation of formalized procedures for the realization of practical ends. Economic theories and models are technologies in this sense, in that they aim to calculate and predict outputs based on inputs, aiming to realize the most efficient and effective input-output function. The ecologization of economic theory implies that the notions of efficiency and effectiveness are modified by introducing new variables that refer to natural capital.

Environmental efficiency is indeed the new goal for technology, including the technologies of economics and management science. This efficiency is to be achieved while preserving the cherished values of modernity as much as is possible. The overall system of which the institutions of technology and economy are a part, as well as most of the basic principles of these two institutions, is to remain intact. The increased environmental efficiency and ecological soundness of products produced by a more ecological industry, under conditions of a more ecological economic system, is then to guarantee sustainable patterns of consumption. Serious reform of current systems of consumption and correlated social institutions need not be pursued then. It is not surprising, then, that volume control and the reform of current lifestyles and consumption patterns are not pursued as serious options within the project of ecological modernization. The promise of ecological modernization is that serious reform in these areas will not be necessary, a promise that makes a happy fit with the modernist ideal of economic growth and the ideals of autonomy, freedom, and quality of life that have become embodied in the consumer lifestyle.

A potential embarrassment for the project of ecological modernization may be its insistence that the ideal of unlimited economic growth is compatible with sustainable development. The modernist ideal of economic growth seems to conflict with ecological principles that appear to support the idea of limits to growth (Meadows et al. 1972, 1991; Daly and Cobb 1990). The apparent conflict lies in the fact that economic growth appears to imply an increase in the consumption of natural resources. In response to this problem, some economic theorists have attempted to delineate a conception of economic growth that does not imply increasing consumption of natural resources. For example, Goodland and Ledec (1993) argue that economic growth (as measured by gross national product or a related index) is in principle not related to increasing consumption of natural resources, and may therefore be free of any natural limits. Goodland and Ledec recognize limits to growth in consumption of natural resources, but argue that "growth in economic output may not be similarly constrained, since innovation may continue to find ways to squeeze more 'value added' from a natural resource bundle." They conclude that "governments concerned with long-term sustainability need not seek to limit growth in economic output, so long as they seek to stabilize aggregate natural resource consumption" (ibid., 252).

This view explains how it is possible that, historically, the idea of sustainable development has been tied to economic growth. It explains, for example, how in the Brundtland report—the very report responsible for popularizing the idea of sustainable development—it can be claimed that economic growth of 3–4 percent per annum for industrialized nations and 5–6 percent for developing nations is desirable (WCED 1987, 50), and need not lead to a further loss of natural resources (ibid., 52). Often it is even claimed that economic growth benefits the environment (and economic stagnation hurts it), because poverty and environmental problems are intrinsically related and because economic growth is necessary to finance the costs of ecological modernization.

Limitations of Ecological Modernization

Critiques of EM come in two kinds: critiques of EM as a successful strategy for sustainability on its own terms and critiques that are really not targeted at EM itself but at the institutions of modernity that it sets out to preserve.

In this section, my focus will not be on critiques of the latter type, but only on the likelihood of EM's success as a strategy for sustainable development. Since the success of EM is ultimately an empirical issue, one may point to its success, or lack thereof, in curbing emissions and environmental degradation as evidence for or against it. However, the success of EM is currently difficult to assess on empirical grounds. Proponents may point to the fact that many countries have succeeded in curbing certain forms of environmental degradation or certain emissions, in increasing energy efficiency, or in making early steps in transitioning from fossil fuels to renewable energy. However, opponents may point to studies that show that global emissions and environmental degradation are up, and that almost 40 years of efforts according to the approach of EM have not brought society close to sustainable development.

Rather than exploring controversial empirical arguments for and against EM, I will here consider two theoretical arguments against it: the argument against technological neutrality and the argument against unlimited economic growth. These arguments challenge fundamental assumptions that are inherent in EM and in the underlying beliefs of modernity.

The argument against technological neutrality points to a flaw in the project of ecological modernization: its retention of an instrumentalist, Enlightenment conception of technology. It is an assumption of the ecological modernization project that the environmental crisis can be solved through mostly technological means, and that a technological reform enables a controlled ecological modernization of production systems that makes them ecologically sound while retaining high output. Ecological modernization has the characteristics of a technological fix: the solution of a complex social problem through technological as opposed to addressing its root causes. This faith in a technological fix for environmental problems can be criticized because critiques of instrumentalist conceptions of technology have taught us that technological solutions often have unwanted and unexpected side effects, and a technological solution may simply not be possible for any social problem. The particular side effects of technological reform within the project of ecological modernization are likely, I argue, to undermine this very project as a control strategy for sustainable development.

The idea that technologies are not neutral and standardly have unanticipated and undesirable side effects is, of course, not new in the philosophy of technology. Important, however, are the details of how this idea applies to the project of ecological modernization and works to undermine it. The most fundamental reform strategy of ecological modernization was identified earlier in this chapter as the structural technological reform of production systems, involving such strategies as integral chain management and quality improvement. Now let us consider the strategy of integral chain management. In this control strategy, the aim is to modify production processes and corresponding products such that material cycles are created that are closed off as much as possible, with a minimum of emissions and waste streams. Central to this strategy are the recycling of used-up products and of wastes generated in production, the use of renewable raw materials, and, when recycling is not an option, the use of biodegradable product materials. The optimism that sustainable production processes based on the principles of integral chain management will generally be possible may, however, turn out to be unjustified. Consider, first, the implications of a move toward the use of renewable and biodegradable materials in integral chain management. Smits (1996) explains how the use of such materials may fail to yield a more sustainable production process. She considers a hypothetical case in which most future polymers (plastics) are produced from renewable materials such as corn starch rather than from nonrenewable resources such as petroleum. "Considering the current heavy demand for polymers," Smits writes, "such a development would necessitate a considerable increase in the scale and intensity of agriculture. How much farming land, pesticides, acidification and erosion of the soil, damage to landscape or expulsion of local inhabitants would be needed to fulfill the demand for polymers?" (ibid., 218). Extensive product recycling in integral chain management may have similar side effects. As Smits explains, a recycling economy would require added transportation of wastes and waste selection and reprocessing, processes that are energy-intensive. "What is the use of almost closed material cycles," Smits writes, "if these cycles themselves turn around faster and faster? Environmental policy aimed at sustainable development by way of integral chain management could possibly choke in its own goals." (ibid., 219)

The argument against unlimited growth points to a second flaw in the project of ecological modernization: its attempt to reconcile the ideal of sustainable development with the ideal of unlimited economic growth. As was explained in the previous section, the defense for the compatibility of these two ideals rests on the assumption that increased environmental efficiency of technologies will offset expected increases in environmental degradation. New technologies (such as micro-electronics and genetic engineering), new materials, and new environmental technologies and procedures (such as integral chain management) are thought to be instrumental in attaining increases in efficiency. They will enable the extraction of more and more economic activity from the same stock of natural resources, while stabilizing pollution and waste streams. The consequences of the use of these technologies are increasing *dematerialization* (that is, the use of less or lighter materials for technologies that yield the same functionality; see Herman et al. 1989), more durable goods, less waste, waste that tends to be more biodegradable, less or less harmful emissions, and an increase in energy efficiency.

Obviously, these developments may lead to the production of more environmentally efficient technologies. However, two objections may be made against the idea that the promise of increased environmental efficiency of technologies allows for economic growth without increased damage to the environment.

The first objection entails a historical argument: Promises that new technologies would help solve the environmental crisis have been made since the 1970s; however, these promises have not been fulfilled, because increases in environmental efficiency have tended to have been offset by economic growth. When nations desire to keep up economic growth of 3–4 percent per annum, the environmental efficiency of technologies has to increase by at least that amount each year. Future developments may make that possible, but past developments have not provided any reasons for optimism.

A second, more principled objection is that there appear to be limits to the increases in environmental efficiency that are attainable. Dematerialization, for example, clearly has limits, because repeated reduction in the mass of many artifacts would lead either to losses in functionality or to losses in durability or safety. Moreover, as has already been pointed out, many new environmental technologies may have environmental side effects that ultimately make them unsustainable. As has also been noted, increases in environmental efficiency achieved through new, renewable, and biodegradable technologies and progress toward a recycling economy may be limited. It can be concluded, then, that the hypothesis that unlimited increases in environmental efficiency are possible rests again on an unjustified faith in technology to fix problems. The hypothesis that the efficiency gains of ecological modernization will outpace growth in consumption is without substantiation and therefore little more than a gamble.

However, proponents of ecological modernization may argue that, although the reform of industry may not give us sustainable development, the ecologization or greening of the economy will. By subsidizing and supporting economic activity that is environmentally friendly and by taxing and penalizing economic activity that is environmentally harmful, it may be argued, the economy, and thereby society, will eventually be made to conform to the principles of sustainable development. The greening of the economy does not only help the greening of production and distribution, but also the greening of consumption. Through economic incentives, consumers will eventually purchase environmentally friendly products because they are less expensive than environmentally harmful ones, and they will be stimulated to conserve energy and recycle. In this way they will, through economic incentives, be stimulated to reduce their ecological "footprint" and adopt sustainable patterns of consumption.

However, the economic aspect of ecological modernization appears to have the same problems as its technological aspect in that it assumes that the efficiency gains of ecological modernization will outpace growth in consumption. Economic incentives may stimulate people to purchase environmentally friendly products, such as energy-efficient lamps and hybrid cars; however, with the growth of the economy and concomitant increases in wealth and income, people will have larger houses, more lamps, and more automobiles, so that increases in consumption may well offset the environmental gains made in the greening of consumer products and particular consumption processes.

Sustainable Consumption as a Condition for Sustainability

If ecological modernization does not yield sustainability, which alternative environmental strategies do? Any such strategies will have to abandon the classical modernist ideal of economic growth and the belief in a largely technological solution to the environmental crisis. This is not to say that any ideal of economic growth will have to be abandoned, let alone the modernist ideal of progress. What will have to be abandoned is the idea of unlimited economic growth attained by continuous growth in the quantity and economic value of goods available in different economic sectors (mining, agriculture, forestry, fishery, automotive, electrical, energy, metallurgical, textiles and clothing, consumer goods, and so on). Significant sustainable growth could still be possible through intense processes of ecological modernization. However, even if it is possible, such growth will have limits, and these limits may have already been reached.

Abandoning the classical ideal of economic growth also does not imply giving up on the ideal of progress. The Enlightenment ideal of progress was initially not formulated as a belief in economic growth, but rather as a belief that science, technology, and reason could better the human condition and improve the quality of life. Only much later did quality of life become linked to economic prosperity, which happened in classical economics through the notion of utility. In recent decades, however, a measure of quality of life, or well-being, has been developed separately from economics in studies of happiness, well-being, and quality of life (Kahneman, Diener, and Schwartz 1999).

The study of well-being has also affected economics, in which the correlation between economic processes and individual well-being has become a topic of study (Frey and Stutzer 2002). "Happiness economics" considers how economic factors such as income, wealth, unemployment, and social security, as well as social and institutional factors such as freedom, relationships, and good governance, affect individual well-being. Some economists have even gone so far as to argue that happiness should become the new metric of economics, replacing monetary value or preferences (ordinal utility) as the values that economics aims to optimize. On this conception of economics, economic policies and public policies should not aim to maximize gross domestic product but should instead aim to maximize gross national happiness, as measured by some happiness index.

In part as a result of these efforts, happiness indices and quality-of-life indices have taken on a major role in public policy in the past twenty years. The Economist Intelligence Unit's quality-of-life index, the UN's Human Development Index, and Gallup's global well-being survey are used to measure and compare happiness or quality of life within nations, cities, or regions. Some countries now are using happiness indices as a supplement or an alternative to gross domestic product (GDP) as a measure of progress; some have even begun to use them to guide national policies. Bhutan is the first country that has decided to measure its progress in terms of gross national happiness (GNH) rather than GDP, using sophisticated surveys to measure the population's level of well-being. Other countries that are using or considering GNH indices include Thailand, China, Australia, Canada, France, and the United Kingdom.

A workable strategy for sustainability should, I claim, replace the ideal of unlimited economic growth with an ideal of limited economic growth within ecological boundaries, and should divorce the notion of progress from the notion of economic growth and refocus it to mean the advancement of well-being in human societies. The major problem for any such strategy is that in current industrial societies the ideal of well-being is itself still strongly conditioned by the ideals of economic success and high levels of consumption of goods and services. Consumerism—the social and economic practice and ideal of consuming ever-increasing quantities and services so as to enhance one's quality of life—is the major obstacle. In a workable strategy for sustainability, a reform of systems of consumption that includes a partial or complete abandonment of consumerism will be necessary. This implies that the strong focus of EM on a reform of systems of production will be supplemented with a strong reform of systems of consumption.

Although ecological modernization has traditionally focused on the reform of production, since the since the 1990s some conceptions of EM have included the reform of consumption (Spaargaren and van Vliet 2000; Spaargaren 2003). In spite of the criticism in the previous section, couldn't EM therefore be a plausible candidate strategy for sustainable development if it also includes a reform of consumption?

EM's position on the reform of consumption is that a restructuring of consumption should not assume that consumption should be limited, but rather that it should be restructured along ecological lines. Such a restructuring could sometimes mean that less is consumed, but it does not involve overall downsizing or limits to consumption (Mol and Spaargaren 2010). The challenge, as EM has it, is to try to restructure consumption patterns and lifestyles through sociotechnical innovations so as to make them more sustainable "without the need to abandon the existing high quality levels of modern consumption" (Spaargaren 2003, 697).

EM's approach to the greening of consumption has, however, received much less attention so far than the topic of production; thus, in practice, EM's emphasis is still strongly on the reform of production processes. A more fundamental problem is that by not abandoning the overall idea of limitless consumption (and, consequently, of limitless growth), and by not advocating an overall downsizing of consumption, EM embraces the ideals of unlimited growth and consumerism that I just criticized. The belief that a sustainable ecological restructuring of consumption is possible without overall limits and downsizing of consumption is not well grounded, and therefore EM's approach to the ecological restructuring of consumption is not likely to succeed.

What is needed instead is an approach to consumption that gives it a central place in a strategy for sustainable development and holds that a reform of consumption implies an abandonment of consumerism and limitless growth. Such an approach will have to be different from the largely technocratic approach of EM that focuses on technological and economic reforms. It will require cultural reform and a reform of values regarding the quality of life and its dependence (or lack of dependence) on high levels of consumption and economic growth. As a result of such reform, people will be prepared to consume less overall and to avoid or minimize those consumptive practices that are unsustainable.

There is increasing scientific evidence that the correlation between well-being and high levels of consumption is weak at best. Since the 1970s, studies have consistently shown that people in high-income countries are not significantly happier than people in low-income countries (at least, those whose incomes are above a certain threshold), and that increases in income above this threshold did not seem to yield significant increases in happiness. This has been called the paradox of happiness (Easterlin 1974). Studies have also shown that people whose values center on the accumulation of wealth or material possessions are at a greater risk of being unhappy, anxious, and depressed whether or not they are successful in such accumulation (Kasser 2002). Many authors have argued that quality of life does not derive from affluence, but from the experience of mental and bodily engagement and connectedness with one's surroundings that is gained through meaningful interaction with one's social and physical environment (Seligman 2002; Borgmann 1984).

How realistic is it that a cultural reform of consumption will be possible in the near future? Aren't people so caught up in consumerism that a radical change in practices will be impossible? I believe there is some reason for optimism. Some studies show that in Western countries a shift has been taking place since the 1970s from "modern" values that center on economic accumulation and social status to "postmodern" values such as freedom, self-expression, and quality of life (Inglehart 1997). Other studies show a decrease in consumerist attitudes in younger generations (Parker, Haytko, and Hermans 2010).

Studies have also shown that in recent decades consumption has changed from a means of meeting material needs to a means of creating personal identity (Hamilton 2010). This development may make change in consumption patterns easier to achieve, since a decrease in the volume of consumption may be easier to accept if it requires changes in how people construct their identities than if it implies that certain needs no longer will be satisfied. A growing awareness of environmental problems among consumers, together with changes in values that undo the perceived connection between quality of life and high levels of consumption, might yield new identities and lifestyles that would emphasize sustainability over consumption.

In recent decades, several social and intellectual movements have emerged that aim to transform consumption to make it more sustainable.

Ethical consumerism (Harrison, Newholm, and Shaw 2005) is perhaps the best known of these movements. It is a form of consumer activism that involves the intent to purchase products that have been produced ethically and that are not harmful to the environment or to society, and to avoid or boycott those that do not meet those standards. Environmental considerations have traditionally been central to ethical consumerism. However, ethical consumerism does not directly challenge consumerism itself or the idea of unlimited economic growth. It is therefore not necessarily a strategy that will yield the radical reform of consumption processes that is needed to make them sustainable.

Voluntary simplicity, or *simple living* (Etzioni 1998), a movement with American origins, was not born out of ethical concerns, but out of a desire to enhance the quality of life for individuals. It tends to involve fewer possessions, less consumption, less work time, increased self-sufficiency, a simplified diet, and being satisfied with what one has rather than wanting more. It sometimes also includes the use of simpler technology or even a complete renunciation of technology. Voluntary simplicity can be understood as a radical form of *downshifting* (Schor 1998), a broader social trend aimed toward finding an escape from the stress that comes from economic pursuit. Downshifting implies finding a better balance between leisure and work, accumulating fewer possessions, and focusing on personal fulfillment and the building of relationships rather than on economic success. Both voluntary simplicity and downshifting appear to be compatible with the kind of reform of consumption that is needed for sustainability.

The more recent *degrowth* movement, which has its roots in Southern Europe, is a social, political, and economic movement which holds that attaining a sustainable society will require imposing limits on growth, that both production and consumption should be reduced (Demaria et al. 2013), that a decrease in consumption need not result in a decrease in well-being, and that people should aim to maximize happiness and well-being through means other than consumption, such as art, music, family, culture, and community. Most degrowthers are also anti-capitalist, holding that capitalist economies unavoidably promote unlimited growth, consumerism, and greed at the expense of solidarity and justice.

I do not want to argue that any of these movements presents the perfect path toward the ecological reform of consumption and toward a sustainable society. However, their existence is evidence of a growing interest in addressing environmental problems through a reform of consumer culture. But what will be the place of technology in such a society, and what role can technology play in making a change in consumer culture possible?

Technologies for Sustainable Lifestyles

Most consumer goods, including cars, computers and other electronic devices, furniture, and clothing, are the results of industrial production processes and thus qualify as technological products. The question I aim to answer in this section is how the market for consumer goods will have to be reformed to allow for sustainable consumption. Sustainable consumption is evidently not dependent on supply alone, but also on demand and on usage patterns. But my focus will, at least initially, be on the supply side.

An ecological restructuring of the market for consumer goods will have to involve two types of reform: (1) the introduction of consumer products that are themselves sustainable and that promote sustainable behaviors and lifestyles and (2) the reduction or elimination of consumer products that are unsustainable and that support unsustainable behaviors and lifestyles.[1]

Producers, consumers, regulators, and civil-society organizations will all have roles to play in attaining these two types of reform. Producers, as developers of new products, ought to have a leading role in the first type of reform. Ways in which their products can contribute to sustainable consumption include the following:

- use of sustainable materials (e.g., biodegradable plastics, recyclable metals) and sustainable, renewable energy sources (e.g., devices that run on solar energy or on "green" batteries)
- designing for energy efficiency in products that consume energy
- using durable materials to make long-lasting products that can be repaired or upgraded
- adopting product life-cycle approaches in which a product's total environmental impact is accounted for, from raw materials to production, distribution, consumer use, and disposal
- designing products that impede or discourage unsustainable behaviors and lifestyles and encourage conservation (e.g., showers that switch off after five minutes of use)
- designing products that support or require sustainable behaviors and lifestyles (e.g., products that make bicycling more attractive)

Of these reforms, the first four are currently well known in ecological design. The last two, however, have not yet received as much attention. They both rest on the idea that products can be designed so as to steer or influence behaviors, attitudes, and lifestyles. Although that idea is not universally accepted, in recent years several approaches to design have been developed that incorporate it.

One class of approaches goes under the name *persuasive technology* or *persuasive design* (Fogg 2003; Wendel 2013). Persuasive technology is technology that is designed to change attitudes or behaviors of the users through persuasion and social influence rather than coercion. Most of its applications are in the design of computing technologies, which draws heavily from experimental psychology and from studies of human–computer interaction. Persuasive technologies may stimulate certain actions by making them easier, more attractive, or more pleasant to perform. Persuasive technologies can send emails to encourage people to take certain actions, they can create experiences that allow for behaviors to be explored, rehearsed or empathized with, and they can help communicate social approval or disapproval for certain behaviors. Environmental design aims to persuade users to engage in more environmentally sound behaviors. For example, some cars now have fuel-economy meters that encourage people to drive at more economical speeds. There are educational computer games that provide players with information about the relationship between various behaviors and household energy consumption, thus encouraging more sustainable behaviors.

Persuasive design is different from design that requires or prevents certain behaviors. Cars with speed delimiters or appliances that automatically switch to solar power if there is enough sunlight may be perceived as "user-unfriendly" and may meet with resistance.

Another class of design approaches could go under the name *design for well-being* (Brey 2014). These are approaches to design that aim to enhance the well-being of users. As I argue in the book just cited, a number of approaches to design for well-being have emerged in recent years, including life-based design, emotional design, and design influenced by "positive psychology." Ruitenberg and Desmet (2012), for example, have developed a positive-psychology approach to product design that focuses on long-term life satisfaction rather than short-term experiences or emotions. Their designs are intended to support meaningful activities that develop users' skills and talents, that are rooted in users' values, that contribute to a greater good, and that are rewarding and enjoyable in themselves. The design method includes visualizing meaningful activities and then designing products that enable or inspire people to engage in those activities.

Many approaches to design for well-being take the psychological literature on well-being seriously. Such approaches focus less on short-term pleasures and consumer experiences and more on supporting meaningful experiences, social relationships, engagement with one's physical and social environment, self-improvement, and long-term life satisfaction. By supporting and fostering non-consumerist behaviors, values, and lifestyles, such approaches may support a transition to more sustainable lifestyles.

Another way in which producers can support sustainable consumption does not involve design. Producers are normally also involved in marketing their products. Much of present-day marketing is concerned not with promoting a product's qualities and benefits, but with promoting a lifestyle in which the product is positioned—for example, a luxurious, or healthy, or achievement-oriented, or experience-oriented lifestyle. Such lifestyle marketing can also be used to promote sustainable lifestyles into which the marketed product fits. Thus, it may be practicable to market products in a direct and overt manner for their fit with a sustainable lifestyle. Since such marketing probably does not yet have appeal for large portions of the population, sustainability could also be marketed indirectly by promoting lifestyles for personal well-being that also happen to support sustainability. In this way, marketing can help individuals construct and strengthen identities that include ecological sensibility and can stimulate the sustainable use of products.

Conclusion

In this chapter, I have investigated the importance of sustainable technology as part of a strategy toward sustainable development. I have analyzed how technology should be developed for it to become sustainable. I have described the current dominant strategy for sustainable development as ecological modernization, an environmental control strategy that aims a greening production processes, and to a lesser extent also consumption processes, in a way that leaves existing institutions and practices intact as much as possible and that focuses on the ecological transformation of technology and production. I have argued, on both theoretical and empirical grounds, that this approach is not likely to result in sustainable development. A more fundamental reform of some of the systems and underlying values and beliefs of modernity will be needed. Most centrally, I have argued, fundamental reform is needed in patterns of consumption, in modern Western lifestyles, and in the values and beliefs that underlie such lifestyles.

I have argued that there is an approaching consensus in empirical studies of well-being and happiness that there is a weak correlation at best between well-being and high levels of consumption, and that consumerist lifestyles may actually make people unhappy. In addition, there appears to be an increasing receptiveness among the public to embrace new lifestyles that move beyond consumerism and materialism. I have argued that, because of the need for reform of consumptive practices, the development of sustainable technology should not focus only on ecological principles in production technology and eco-efficiency, but also on supporting sustainable consumptive practices and lifestyles. I have presented ways in which such an greening of technologies for consumption may be realized.

It should not be thought, however, that the redesign of technologies to promote sustainable consumption will be sufficient in itself to engender sustainable systems of consumption. The idea that this is possible amounts to another belief in a technological fix, this time by the "social engineering" of lifestyles and patterns of consumption through a reform of technology. As an isolated strategy, such reform will fail because consumer preferences and market competition by other technologies are likely to lead to a rejection of redesigned technologies by most consumers in favor of technologies that are less sustainable but make a better fit with their ideal of the good life. Technological reform will certainly be of great help in the move toward sustainable patterns of consumption. However, such reform should be seen as part of a comprehensive strategy for sustainable consumption that also includes social and economic incentives and public debates about values, lifestyles, the quality of life, and the future of the planet.

Note

1. Although proponents of EM are likely to argue that both of these reforms are consistent with EM, it is fair to say that EM emphasizes technological solutions that are minimally intrusive to consumerist lifestyles. The difference between EM and the position I advocate is ultimately one of degree: How radical will the reform of systems of consumption be? The position that I advocate would entail a greater emphasis than in EM on reducing consumption, on limiting and eliminating unsustainable products, and on introducing products that support or require serious reorientations in behaviors and lifestyles.

References

Borgmann, A. 1984. *Technology and the Character of Contemporary Life*. University of Chicago Press.

Brey, P. 1997. Sustainable technology and the limits of ecological modernization. *Ludus Vitalis. Revista De Filosofia de Las Ciencias de la Vida / Journal of Philosophy of Life Sciences* 7 (12): 17–30.

Brey, P. 2014. Design for the value of human well-being. In *Handbook of Ethics, Value and Technological Design*, ed. J. van den Hoven, I. van de Poel, and P. Vermaas. Springer.

Daly, H., and J. Cobb. 1990. *For the Common Good*. Greenprint.

Demaria, F., F. Schneider, F. Sekulova, and J. Martinez-Alier. 2013. What is degrowth? From an activist slogan to a social movement. *Environmental Values* 21 (2): 191–215.

Easterlin, R. 1974. Does economic growth improve the human lot? Some empirical evidence. In *Nations and Households in Economic Growth: Essays in Honor of Moses Abramovitz*, ed. P. David and M. Reder. Academic Press.

Etzioni, A. 1998. Voluntary simplicity characterization, select psychological implications, and societal consequences. *Journal of Economic Psychology* 19: 619–643.

Fogg, B. 2003. *Persuasive Technology: Using Computers to Change What We Think and Do*. Morgan Kaufman.

Frey, B., and A. Stutzer. 2002. *Happiness and Economics*. Princeton University Press.

Goodland, R., and G. Ledec. 1993. Neoclassical economics and principles of sustainable development. In *Environmental Ethics: Divergence and Convergence*, ed. S. Armstrong and R. Botzler. McGraw-Hill.

Hamilton, C. 2010. Consumerism, self-creation and prospects for a new ecological consciousness. *Journal of Cleaner Production* 18 (6): 571–575.

Harrison, R., T. Newholm, and D. Shaw. 2005. *The Ethical Consumer*. SAGE.

Herman, R., S. Ardekani, and J. Ausubel. 1989. Dematerialization. In *Technology and Environment*, ed. J. Ausubel and H. Sladovich. National Academy Press.

Huber, J. 1982. *Die Verlorene Unschuld der Ökologie. Neue Technologien und superindustriele Entwicklung*. Fisher.

Illich, I. 1973. *Tools for Conviviality*. Harper and Row.

Inglehart, R. 1997. *Modernization and Postmodernization: Cultural, Economic, and Political Change in 43 Societies*. Princeton University Press.

Jacobs, M. 1991. *The Green Economy*. Pluto.

Kahneman, D., E. Diener, and N. Schwarz, eds. 1999. *Well-Being: Foundations of Hedonic Psychology*. Russell Sage Foundation Press.

Kasser, T. 2002. *The High Price of Materialism*. MIT Press.

Meadows, D. H., D. L. Meadows, and J. Randers. 1991. *Beyond the Limits: Confronting Global Collapse; Envisioning a Sustainable Future*. Earthscan.

Meadows, D. H., D. L. Meadows, J. Randers, and W. Behrens. 1972. *The Limits to Growth*. Universe Books.

Mol, A. 1995. *The Refinement of Production. Ecological Modernization Theory and the Chemical Industry*. Van Arkel.

Mol, A., D. Sonnenfeld, and G. Spaargaren, eds. 2009. *The Ecological Modernisation Reader: Environmental Reform in Theory and Practice*. Routledge.

Mol, A., and G. Spaargaren. 2000. Ecological modernisation theory in debate: A review. *Environmental Politics* 9: 17–49.

Mol, A., and G. Spaargaren. 2010. Ecological modernization and consumption: A reply. *Society & Natural Resources* 17 (3): 261–265.

Parker, Richard S., Diana L. Haytko, and Charles M. Hermans. 2010. The perception of materialism in a global market: A comparison of younger Chinese and US Consumers. *Journal of International Business and Cultural Studies* 3 (May): 1–13.

Ruitenberg, H., and P. Desmet. 2012. Design thinking in positive psychology: The development of a product-service combination that stimulates happiness-enhancing activities. In *Out of Control: Proceedings of the 8th International Conference on Design and Emotion*, ed. J. Brassett, P. Hekkert, G. Ludden, M. Malpass, and J. McDonnell. Central Saint Martins College of Art and Design.

Schor, J. 1998. *The Overspent American. Upscaling, Downshifting, and the New Consumer*. Basic Books.

Schumacher, E. 1973. *Small Is Beautiful: A Study of Economics As If People Mattered*. Blond & Briggs.

Seligman, M. 2002. *Authentic Happiness: Using the New Positive Psychology to Realize Your Potential for Lasting Fulfillment*. Free Press.

Simonis, U. 1989. Ecological modernization of industrial society: Three strategic elements. *International Social Science Journal* 121: 347–361.

Smits, M. 1996. The unsustainability of sustainable technology. In *Polymer Products and Waste Management: A Multidisciplinary Approach*, ed. M. Smits. International Books.

Spaargaren, G. 2003. Sustainable consumption: A theoretical and environmental policy perspective. *Society & Natural Resources* 16: 687–701.

Spaargaren, G., and A. Mol. 1992. Sociology, environment and modernity. Ecological modernization as a theory of social change. *Society & Natural Resources* 5: 323–344.

Spaargaren, G., and B. van Vliet. 2000. Lifestyles, consumption and the environment: The ecological modernization of domestic consumption. *Environmental Politics* 9: 50–77.

WCED (World Commission on Environment and Development). 1987. *Our Common Future.* Oxford University Press.

Wendel, S. 2013. *Designing for Behavior Change. Applying Psychology and Behavioral Economics*. O'Reilly Media.

12 Sustainable Animal Agriculture and Environmental Virtue Ethics

Raymond Anthony

Technology does not merely refer to artifacts that have functional purposes. It is a "form of life" with its own agency, for it shapes, organizes, governs, enables, and limits patterns of human behavior. In the wake of observations made by Latour (1991, 1992), technology is its own edifice, comprising artifacts, techniques, and technical capacities that pattern human activities. However, social critics highlight the fact that technology as a form of life is value-laden, interest-laden, and not ethically neutral (Stump 2006). And yet the meanings of technology are not fixed by designers but essentially up for grabs. There is a kind of ambivalence to technology—social relations change and adjust continually when new technologies permeate the public arena. Feenberg (2003) urges that the public not leave technological affairs in the hands of experts or technocrats. Instead, engagement with technology should involve careful negotiation between technocrats and the public in order to ensure that the design parameters that will be integrated into the lives of individuals and communities reflect broader public values and not be dominated by the motives or values of an exclusive few (ibid.). This form of democratization of technology requires public involvement in technical change (Veak 2006).

Concerns about the modern technological edifice also have bearing on the industrial food complex. We can reasonable ask just how and to what extent the global food system as a modern technological edifice should be controlled or governed through wider public participation. Yet the arguments marshaled in favor of sustainable agri-food systems (in general) and of promoting animals' interests (in particular), by and large, are still usually discussed without reference to the role that technologies play, much less the dominant philosophy of technology that currently motivates how we farm. Many of the influential recent discussions about the welfare of farmed animals have focused narrowly on philosophical arguments concerning the moral status of animals and their membership in the moral community (Singer 1990; Regan 1983, 1991; Midgley 1983, Nussbaum 2004) or on the science of animal welfare, and whether or not the adaptations possessed by farmed animals meet the demands of the production system and whether they are free from physiological, psychological, and physical harm or disease (Fraser 2001; Fraser and Weary 2004). These discussions affect personal choices about the consumption of animals and animal products, with very little consideration of how specifically animal issues are connected to and mediated by any understanding of what technology is and how it influences how we experience things. (See Rollin 1995; Berry 1996, 2009; Hunkel 2000.)

My modest response to the lacuna mentioned above is divided into two sections. In the first, I attempt to link critical reflections on technology from Andrew Feenberg and Albert Borgmann to the currently pervasive view of farmed animals as mere resources or commodities. Employing Feenberg's and Borgmann's analyses of our present relationship with technology, I contend that appreciating the good of animals in their own right is obfuscated by the fact that food production under the industrial model (propelled by the winds of the free-market economy) has become a device and that we are seduced or blinded by "the promise of technology." In the second section, I suggest a virtues-motivated approach to technology as a poignant way to turn from the dominant instrumental philosophy of technology and to encourage a shared form of institutional governance of the industrial food system. I will explain four elements of an approach based on an institutional virtue ethic of care. My proposal calls attention to collective action to be taken by individuals with commitments to be virtuous consumers and by industry decision makers who aspire to promote public-regarding concerns in the food system.

Technology and Commodification

A central project among many recent scholars of technology concerned about its social and political aspects has been to debunk what is referred to as the "essentialist paradigm" in favor of alternative design philosophies. Essentialism perpetuates the notions that technological development follows a single, fixed path of necessary stages and that it is self-directing. The apparently deterministic character of technology and our relationship with it in present-day life, according to Feenberg (1999), has taken hegemonic form. It is deeply woven into the fabric of collective social life, and changing it is no small feat. The technological hegemony of our day found in the industrial paradigm focuses on efficiency. Society's emphasis on efficiency serves "the promise of technology," i.e., the promotion of a culture of convenience, or the disburdenment from laborious engagement with the world so that we may concentrate on other seemingly more worthwhile matters (Borgmann 1984).

The success of the essentialist paradigm is, however, contingent on a particular distribution of socio-political power. Although technology is often experienced as beyond our control, social institutions need not bend to the will of the seeming technical imperative. Technology is inherently social and political. It depends on users, not merely on designers and engineers, for its meaning and normative content. Feenberg (1999) argues that technology is a site of contestation of philosophical alternatives; the apparent essentialism of the techno-political hierarchy in present-day society is not inevitable. It is social through the purposes it serves, and the purposes should not be left only for a technical elite to design upstream, but must also correspond with the contexts or realities of the users downstream. A technology that is sensitive to and reflects a range of public interests, values, and functions is likely to be more sustainable for having withstood the test of public scrutiny.

In the case of the industrial food complex, determination of the shape and ends of the system by a select few over a majority of others has left a sour taste among many consumers concerned about animal welfare, environmental ethics, and human rights. The imposition of intensification as the dominant technology for food production in the twentieth and twenty-first centuries is a product of the mindsets of essentialism and technological determinism. The social policies behind industrial modernization and the food policies of the 1970s that advanced a policy of "cheap and plentiful" show a limited understanding of food and food security that has resulted in harm to animals and in ecological costs (Ilea 2009). Belief in the apparent technical imperative associated with the industrial paradigm has effectively frustrated the development of any alternative goals other than technical efficiency. This has led to numerous deleterious effects and to the growing public outcry that we hear today from various alternative food movements.

Drawing from Borgmann's work on the character of technology (1984, 2006), we begin to see more clearly how this mode of food production to increase agricultural yield per unit input in the form of intensification or agricultural industrialization has not only influenced physical transformations of landscapes, rural development, and migration patterns, but also has altered profoundly our relationship with technology and the nonhuman world. For Borgmann, the aims of culture over-emphasize convenience, or, as he puts it, disburdenment. The central technological ingredient in our culture of convenience is the device. A device is something to be manipulated and controlled; it produces a commodity that is used to bring about some human end in the most efficient way. Devices are disburdening and make no demands on us. They are disposable, replaceable, and anonymous. They relegate us to the status of mere consumers of products who need not know anything about the inner workings or the production histories of the devices themselves. The device paradigm and the ethos of disburdenment perpetuate the widespread reproduction of artifacts that contribute to the idea that, ultimately, efficiency and convenience are sufficient for human flourishing. This is, of course, a false conception that is notoriously dangerous to human freedom.

Borgmann (1984) cautions against being seduced by this "promise of technology." Heavy reliance on artifacts or devices comes at a cost: They disengage us from a richer connection with our world. In bringing about only human ends, devices often conceal the manner in which they do so. For example, in animal production the technical design and imperative of commodification conceals the nature of animals and potentially discourages us from properly caring for humans, as well. It nullifies the Agrarian Ideal, in which closeness to farmed animals contributes to the development of moral sentiments such as sympathy and compassion, and limits how we can treat them. The presence of farmed animals and farms in local communities helped to "gather" certain virtues of care, respect, and self-mastery (Thompson 1993).

In the agri-food system, food production has become a veritable commodity-producing black box, which many of us welcome blindly under the cloak of food security and prosperity. The industrial agriculture complex asserts that the fundamental character of food is a technological artifact or device. Like any device, it contributes to the culture of convenience or disburdenment. (Here I have in mind agricultural policies that perpetuate the policy of "cheap and abundant.") Food production is reduced to creating uniform commodities and convenience items (Thompson 2001, 2008). In our culture of convenience, food production is also socially disburdening, since the demand on our faculties is usually felt only in a grocery store or at mealtime and since food usually is consumed in a hurry or as a chore. Since we do not develop a relationship with the food we eat, it remains impersonal and alien.

Animals as commodities

Though farmed animals have been bought and sold throughout history, they were traditionally raised and used in a manner that was consistent with their natures and their adaptations. But with the above-mentioned design philosophy of present-day agri-food complexes, farmed animals have become mere products, nothing but commodities—a manipulated input waiting to be altered into a desired consumable form to a degree that was not been witnessed during the agrarian economy. As commodities, farmed animals are anonymous. They are "absent referents" (Adams 2000), conceptualized and experienced as facsimiles of actual "subjects-of-a-life" (Regan 1983). When conceived and experienced as units of production or commodities, and less as beings with moral status or a good of their own, farmed animals are easily forced into situations for which they lack the requisite adaptations in order to meet the demands of industry. Production practices that handle billions of these absent referents in a year continue to "squeeze [these] round pegs into square holes" (Rollin 1995). While we enjoy the benefits of disburdenment, farmed animals suffer a disproportionate amount of the costs as a result of our industrial technologies. That is to say, consumers have no real relationship with farmed pigs, chickens, and cows. They remain anonymous and thus interestless—abstract cogs that are merely part of the supply chain. Farmed animals in their industrial circumstances further validate the ideas of human self-importance and separation from and dominance of the nonhuman world. Consumers and industry agents who are supportive of this technological system effectively become cogs themselves in a highly mechanized assembly line. This technological edifice keeps contracted farmers mired in debt and alienated from their animals.

On closer inspection of the currently dominant view of technology, the process of commodification of animals can be analyzed in a twofold sense. Farmed animals experience both *institutional commodification* and *technological commodification* (Thompson 2006). In the case of institutional commodification, while there is no direct alteration to the physical nature of the object in question, our treatment of it changes as a result of customs, laws, and practices around the commodity. According to Thompson, institutional commodification is facilitated through social practice around the commodity in question. Institutional commodification is made possible by the legitimization of farming practices that alienate animals from their sentient natures, by the ubiquity of corporate ownership of food commodities, and by the fact that they have become interchangeable objects that are indistinguishable from one another. With the advent of industrial and global agriculture, new actors have emerged in agri-food production, including animal agriculture. Corporate interests and agribusinesses currently dominate the political and economic foodscape and proliferate practices that conceal the nature of food production. The corporate control of food production has spawned a culture of technicians dedicated to the single virtue of efficiency. The centralized and vertical power structure keeps consumers in the dark under the promise of cheapness and abundance.

In the case of technological commodification, the animals themselves are materially transformed. The transformation of the cultural understanding of food animals into their elemental protein parts and the requirement that food be served fast and functional reflect the institutional commodification of animals described above. Technological commodification of animals also occurs in order to meet economic expectations, farming practices, and processing-related criteria such as refrigeration and transportation. Farmed animals have been transformed through genetic engineering or bred to conform to the exigencies of concentrated animal-feeding operations in various ways. The sheer numbers of animals forced into existence, raised in confinement, mechanically processed, and exchanged as a commodity has diminished our view of their moral status and of our own.

Environmental Virtue Ethics as a Response to Animal Commodification

Animal production is a central aspect of sustainable food production of and how we feed people. According to the Food and Agricultural Organization, approximately 56 billion terrestrial animals are raised and slaughtered per year in the world (FAO 2008), and it is projected that worldwide farm animal production will double by 2050 (FAO 2006). Furthermore, how we farm animals in the future has critical implications not only for animals themselves but also for human food safety and security, human health and well-being, the distribution of resources, and the environment. It is critical that we reconsider our current philosophy of technology as a feature of our environmental ethic not only to address the animal issues discussed above, but for our own futures as well.

The project to reinstitute the subjectivity of animals as part of reinvigorating our moral relationship with food is not a process to eliminate use of animals entirely, but it is one that is concerned with restoring a respectful attitude toward them. Given the enormity and complexity of the circumstances in which animals are farmed, any such initiative must involve rethinking the governance structure of the food system as an integral part in this reformation. Governance here should be understood as "translation of collective moral intentions [that must meet appropriate moral standards], into effective and accountable institutional actions" (McDonald 2001, 3–4). Alternative forms of animal agriculture that are more humane and sustainable push against the tide of the industrial food system. They reflect the desire for alternate philosophies of agriculture that promote increased decommodification. They also reflect mini-rebellions against surrogate decision making by industry entities and government powers that have inadvertently ushered in an era of living with the technical/functional ideal of efficiency that uncritically puts animals, the planet, and ourselves at risk.

A promising counterbalance to our narrow view of technological rationality is environmental virtue ethics, whose central evaluative concepts are excellences of character since it focuses on living well and cultivating character traits that contribute to flourishing for human beings and the nonhuman world alike (Newton 2003; Sandler and Cafaro 2005). A central tenet of this ethic involves taking ownership for choices that we make, especially in the face of relationships that involve vulnerable or dependent others. The premise of virtue ethics is a conception of the self bound to others both personally and through the various institutions that shape and facilitate life. We are necessarily embedded within human and biocentric communities, i.e., interconnected to and dependent on others through practical and moral nexuses we ought not neglect nurturing.

Virtue ethics within the environmental framework is oriented toward the flourishing of both human and nonhuman individuals and communities. A central advantage of the language of virtue and vice is its richness and depth in confronting the complexity and diversity of the relationships we have with the natural and built environments relative to, for example, the languages of deontological ethics or consequentialism (Sandler 2007). Environmental virtue ethics does not appeal to a one-size-fits-all view. Instead it affords us a cluster of ingredients to respond to our ethical duties, and a dynamic way to discuss and assess a wide array of environmental issues. A virtues-centered approach also implies a pluralistic response to environmental challenges.

Through the lenses of virtue ethics, it is evident that there is an urgent need to reverse the cumulative effects of decades' worth of harms from the industrial food system. Here, an environmental virtues approach can help us build communities and enlighten individuals to orient their lives toward a practicable and an effective environmental ethic. This is a major ingredient in shifting our cultural and philosophical relationship toward technology. Some of the virtues that are conducive to encouraging environmental sustainability include temperance, simplicity, humility attunement, responsiveness, attentiveness, and farsightedness (Cooper 1998).

In the case of our relationship to agriculture, a virtues perspective can help us discern agriculture's role in forming personal moral character and can provide a basis for evaluating policies and transforming technologies. In a traditional account of virtue, a good person strikes a mean position between tendencies of excess and deficiency and aims toward equilibrium after reflecting on and bringing to bear all the facts germane to making an informed choice. Beyond personal morality, the traditional account also acknowledges that our tendencies and abilities to regulate behavior are reflections of the sociocultural and technical environments in which we live. Here, the articulation of ethical norms and standards is likely to call attention to the practices, traditions, and institutions that are characteristic of and valued by a community. A virtues approach challenges the public to create social environments that can give rise to exemplary conduct and encourages people to be poised to pursue a more sustainable, humane, and just world. How, then, should exemplary people act toward a fractured food system and an efficiency-oriented philosophy of technology that undermines the well-being of human beings and animals?

In order to address institutional and technological commodification and concerns about alienation, proprietorship, and uniformity, I propose a virtue of care or caretaking within the framework of an environmental virtue ethics.

Agricultural ways of life have figured prominently in some of the most influential articulations of virtues and vices. For example, when Thomas Jefferson praises farmers he reminds us how farming systems that are committed to citizenship virtues and community solidarity have a profound influence on the strength of democracy in the United States. As part of developing a deeper connection to the place where they live, today's consumers, like smallholding farmers before them, are bound by certain ethical values that they acquire in virtue of being beneficiaries of the industrial food system.

Agricultural virtues, like environmental virtues, are proper dispositions or character traits for human beings to have regarding their interactions and relationships with agriculture, farmed animals, and food. The virtuous person is disposed to respond to farmed animals in an excellent way and to resist reducing sentient beings to mere commodities or mere relative goods.

Here, to care adequately for someone or something with whom we have a relationship or with whom we are situated is a quality of the morally good person or society. The ethic of care starts with an orientation of engagement with the concrete, the local, and the particular. Within this tradition, there are four elements that provide a good starting point for developing a framework of caretaking that can serve to counter our existing philosophy of technology of disburdenment, which has concealed the plight of animals. The following descriptions of these elements are adapted from Tronto 1993 and from Little 1988:

- Attentiveness, which involves being cognizant of what is going on in food production and paying heed to the plight of animals and how our actions influence their welfare. Attentiveness is a disposition to be mindful and an expression of how the world ought to be and what is good (as a reflection of our values). It scorns being mechanical or rote or unthinking in our interactions with others who demand our moral sensitivity.
- Responsibility, which involves the recognition that there is a need to perform certain caretaking functions as a result of our consumerism of some needed or wanted product. The basis of this is found in gratitude to others and in humility for being the recipient of goods produced for our benefit. The desire to minimize the deleterious impacts of our behavior on others flows from our interdependency and our indebtedness to those who bring us a good.
- Competence, which involves discharging one's caring responsibilities in ways that actually bring about good welfare for the ones cared for. Here, along with being attentive to their roles as participants in the system, caretakers can also appreciate the consequences of remote actions performed downstream for others in the system.
- Responsiveness, which involves vigilance of the dependency and vulnerability of those in our charge or the system which we support, being alert to the possibilities of negligence, abuse, or incompetence, and acting accordingly to rectify deficits.

Owing to the nature of the food system, the major actors to whom these virtues are addressed are consumers and industry agents with influence over the technical design—that is, industry technocrats of the food system.

Environmental Virtue Ethics of Care (EVEC) and Consumers

A values-driven agri-food system is defined roughly as the vision of the good life with and for others through humane and fair food-production practices. Currently, the public's lack of appreciation for both traditional and conventional production practices serve as a central barrier for engaging with food in more meaningful ways. By being more attentive to how our choices and obliging technologies may be oppressive to others, we can begin to rebuild the severed connection to food production and farmed animals. Recognizing technology's ability and tendency to shape human behavior and alter values in ways that conceal our responsibilities and the moral subjectivity of the other is a crucial first step in overcoming the ways in which technology contributes to our shortsightedness with regard to food. Attentiveness or mindfulness can encourage more direct public participation in the food system. An attentive citizen-consumer considers the ethical and social impacts of the animal products that she buys and acts upon relevant information by buying animal-friendly products. Attentive consumers limit the likelihood that they will intentionally contribute to the suffering and injustice of others. They are self-restrained and do not overconsume. More importantly, attentive consumers, as part of a collective voice, are also active and thus create meaning through choices that may express non-superficial values about food.

An ethically minded consumer is likely to confront her value commitments and the consequences of her actions from being wedded to the desire for disburdenment. For example, one urgent task is to weigh in on whether genetic engineering promotes a morally virtuous agriculture and food paradigm. Here, consumers must not only be competently informed about the issues; in addition, they bear the responsibility of expressing their values through their actions in political or economic forums where the shape of the technologies can be influenced.

Consumers must own up to how they affect farmers and animals and their complicit behavior through the market economy. Successfully addressing animal issues is a function of our relationship with technology that requires long-term, sustainable changes in the way we choose to live. Animal issues not only challenge our ethics, technology, and politics; they also have the potential to improve the discourse about what we are doing to ourselves—and to our futures if we continue to pursue mindless disburdenment from ethical life. Though it requires much dedication, virtuously minded consumers can reconstruct the links between food production and consumption on their own in local avenues. It is necessary, however, for citizen-consumers to partner with corporate designers and policy makers if advances in non-superficial values-based food production are to be realized.

EVEC and Industry Technocrats

Consumers cannot, however, reform the global industrial food system on their own. Partnership with policy makers and industry agents will be an integral part of any successful transformation of consumer behavior. Businesses, too, have important roles to play in promoting these virtues. The food industry is in business for money. Sometimes it is in the industry's interest to slow down the roll-out of an innovation because producing the new products and technologies may undercut profits if they compete with current ones. But companies have to realize that it is in their interest to move faster in the wake of social disquietude. Here, industry agents who have decision-making power in their capacity as designers or executives must step forward into their new roles as advocates of how to best produce food without compromising ethical integrity. How can we accelerate movement toward a sustainable animal agriculture in conjunction with technological innovation?

Industry agents should share aspects of technological choice with others in order to bring about public-regarding innovation in ways that are better for the environment and address other public considerations. Industry agents, in exemplifying attentiveness, for example, will recognize that they have a significant role to play in reconnecting the severed link between the public and food production and in bringing about changes in animal welfare. Being mindful, they will realize that the industry is taking too long to bring animal-friendly technologies to market and that the industry, under the status quo, cannot be left to its own devices when it comes to innovating for a more sustainable, humane, and fair animal agriculture. Partnership with government agencies must be stepped up so that businesses can share expertise and knowledge about innovations more openly and so that realistic targets can be set, best-practice policies can be put in place, and regulation can be set.

By some estimates, it takes about 20–40 years for new technologies to get widespread use. That is too slow for the billions of animals that continue to live in inhumane conditions. In expressing responsibility as major actors in the system, industrial designers should realize that it is important to find ways to speed up technological change, to promote opportunities for the public to exercise certain democratic rights, and to make it easier for producers and consumers to behave more ethically in the food system. Here, responsibility invites informational transparency so that the public and producers can determine the remote and proximate effects of their actions and make measured responses.

In removing barriers that hinder such participation, industry agents should be open to considering what is wrong with the system from the user's end and not just continue to propose high-tech solutions to maintain an unsustainable and inhumane status quo. Considering the virtues of attentiveness, responsibility, competence, responsiveness together, industry designers must create ways for consumers to inquire and gather information about where their food comes from and the conditions under which it is produced. They must also learn more about the challenges facing farmers and contracted workers, as well as their what their values are. Being attentive and responsible in these ways will help to overcome the tendency to take farm workers for granted (and, in some cases, commodify them). Partnerships are important to help combine goals of efficiency and profitability with improved outcomes for the environment, animals, workers, and the public.

These EVEC virtues are central for industrial designers who exercise discretion over how to innovate, what products to put out in the market, which technological innovations to develop, how products are put together, where to locate facilities, how often to update plants and equipment, and how to promote superior ways of melding business profitability and technological innovation with social and environmental stewardship. A virtuous industry agent in the present case is someone who considers the ethical and social dimensions in a climate of partnership. The ethic proposed here promotes continued critical engagement in the food system, from which it has been absent for far too long. By being attentive and responsive, and by cultivating competence and responsibility, consumers and industry designers alike can resist the somnambulism that pervades our current relationship with food. These virtues can offer a viable way to break the stronghold of a one-dimensional set of values that has, to date, had a disproportionate influence over the global food system.

Lingering concerns

There is no guarantee that EVEC in the variant proposed here will produce the desired ends of a more respectful and sustainable animal agriculture. Arguably, I have presented but one example of a virtue-ethics approach that could work when, in fact, there are other variants of virtue ethics that may imply different relationships with animals and technology. The important question that continues to haunt many in the industrial world concerns the extent to which people are willing or able to unseat the current technological food system—a system that, on the one hand, has given them much and, on the other hand, beckons them to reflect on their self-importance and responsibilities to the rest of the world— and "reseed" it with a different moral orientation that may be costly in the short run. It is my view that grounding an alternative design philosophy for sustainable animal agriculture post-industrialization in EVEC can help us to meet the goals of sustainability and encourage much-needed engagement and progressive change.

References

Adams, C. 2000. *The Sexual Politics of Meat: A Feminist-Vegetarian Critical Theory, Tenth Anniversary Addition*. Continuum.

Berry, W. 1996. *The Unsettling of America: Culture and Agriculture*. University of California Press.

Berry, W. 2009. *Bringing It to the Table: On Farming and Food*. Counterpoint.

Borgmann, A. 1984. *Technology and the Character of Contemporary Life: A Philosophical Inquiry*. University of Chicago Press.

Borgmann, A. 2006. Feenberg and the reform of technology. In *Democratizing Technology: Andrew Feenberg's Critical Theory of Technology*, ed. T. Veak. SUNY Press.

Cooper, D. E. 1998. Intervention, humility and animal integrity. In *Animal Biotechnology and Ethics*, ed. A. Holland and A. Johnson. Chapman and Hall.

FAO 2008. FAO Statistical Database. http://faostat.fao.org

FAO. 2006. Livestock a major threat to environment. http://www.fao.org/newsroom/eb/news/2006/1000448/index.html

Feenberg, A. 1999. *Questioning Technology*. Routledge.

Feenberg, A. 2003. Democratic rationalization. In *Society, Ethics and Technology*, ed. M. Winston and R. Edelbach. Thomson and Wadsworth.

Fraser, D. 2001. Farm animal production: Changing agriculture in a changing culture. *Journal of Applied Animal Welfare Science* 4: 175–190.

Fraser, D., and D. Weary. 2004. Quality of life for farm animals: Linking science, ethics, and animal welfare. In *The Well-being of Farm Animals: Challenges and Solutions*, ed. G. Benson and B. Rollin. Blackwell.

Hunkel, H. O. 2000. *Human Issues in Animal Agriculture*. Texas A&M University Press.

Ilea, R. C. 2009. Intensive livestock farming: Global trends, increased environmental concerns, and ethical solutions. *Journal of Agricultural & Environmental Ethics* 22: 153–167.

Latour, B. 1991. Where are the missing masses? Sociology of a few mundane artifacts. In *Shaping Technology, Building Society: Studies in Sociotechnical Change*, ed. W. E. Bijker and J. Law. MIT Press.

Latour, B. 1992. Technology is society made durable. In *A Sociology of Monsters: Essays on Power, Technology and Domination*, ed. J. Law. Routledge.

Little, P. 1988. *Simone Weil: Waiting on Truth*. St. Martin's Press.

McDonald, M. 2001. Canadian governance of health research involving human subjects: Is anybody minding the store? *Health Law Journal* 9: 1–21.

Midgley, M. 1983. *Animals and Why They Matter*. University of Georgia Press.

Newton, L. 2003. *Ethics and Sustainability: Sustainable Development and the Moral Life*. Prentice-Hall.

Nussbaum, M. 2004. "Beyond Compassion and Humanity": Justice for Nonhuman Animals. In *Animal Rights: Current Debates and New Directions*, ed. C. Sunstein and M. Nussbaum. Oxford University Press.

Pinstrup-Andersen, P., and P. Sandoe, eds. 2007. *Ethics, Hunger and Globalization: In Search of Appropriate Policies*. Springer.

Regan, T. 1991. *Defending Animal Rights*. University of Illinois Press.

Regan, T. 1983. *The Case for Animal Rights*. University of California Press.

Rollin, B. 1995. *Farm Animal Welfare: Social, Bioethical, and Research Issues*. Iowa State University Press.

Sandler, R. 2007. *Character and Environment: A Virtue-Oriented Approach to Environmental Ethics*. Columbia University Press.

Sandler, R., and P. Cafaro, eds. 2005. *Environmental Virtue Ethics*. Rowman and Littlefield.

Singer, P. 1990. *Animal Liberation*, revised edition. Avon Books.

Stump, D. 2006. Rethinking modernity as the construction of technological systems. In *Democratizing Technology: Andrew Feenberg's Critical Theory of Technology*, ed. T. Veak. SUNY Press.

Thompson, P. B. 1993. Animals in the agrarian ideal. *Journal of Agricultural and Environmental Ethics* 6 (Special Supplement 1): 36–49.

Thompson, P. B. 2001. Reshaping conventional agriculture: A North American perspective. *Journal of Agricultural and Environmental Ethics* 14 (2): 217–229.

Thompson, P. B. 2006. Commodification and secondary rationalization. In *Democratizing Technology: Andrew Feenberg's Critical Theory of Technology*, ed. T. Veak. SUNY Press.

Thompson, P. B. 2008. *The Ethics of Intensification: Agricultural Development and Cultural Change*. Springer.

Tronto, J. 1993. *Moral Boundaries*. Routledge/Taylor & Francis.

Veak, T., ed. 2006. *Democratizing Technology: Andrew Feenberg's Critical Theory of Technology*. SUNY Press.

13 Technology, Responsibility, and Meat

Wyatt Galusky

Meat has problems, and many of those problems are intimately tied to technology. Industrialization of meat production has led to increased stresses on animals, on people, and on environments. I confronted this intersection of technology and meat through my own attempts to care for eight chickens, both as a learning opportunity and as a reaction to the problems I perceived to be endemic to our industrialized food system. (See Galusky 2010.) I wanted to be able to take ownership of some small aspect of what fed me, and to shrink the otherwise enormous socio-technical networks that underlay food. What shrank, instead, was the size of my flock, from eight to seven to five to four to one in the course of a few months. Call it a failure of responsibility. In a very real sense, my own failure to be responsible for those chickens is inextricably linked to the larger, more systemic failures to take full responsibility for animals. On the one hand, the impacts of the system prompted my efforts; on the other hand, that system also enabled me to try because so little was personally at stake. If things went wrong, I did not have to scrape the bottom of the barrel (Horowitz 2006); I just had to go back to the supermarket. Reflecting on my failures and my responsibilities led me back to the more systemic failures and responsibilities we have created through more fundamental modifications throughout that system—to animal bodies, to models of work, to communities, economies, and ecologies.

The experience with chickens made me confront some of the central paradoxes attached to food production and technology: industrialization that generates ethical discomfort while providing material comfort; technology that serves as the source of harms and the promise of progress. By making meat more technologically beholden to the standardized logic of machines in raising animals and in processing animal bodies, problems emerge for animals, people, and environments. Yet awareness of the problems associated with industrialized systems of production is often predicated on their success. Industrialization of meat, and our food supply in general, generates enough food to make deprivation an economic, rather than material, condition—a question of access, rather than absence. Many of us, facing a choice regarding what to eat, rather than whether, can now display more sensitivity toward problems associated with how such plenty is achieved because we are not beholden to material scarcity—call it a paradox of abundance. (See Ogle 2013.)

Choices about what to eat expand beyond taste and become expressions of value and demands for solutions. Technology reenters the discussion here, as well, positioned as a potential savior. This is the present-day condition: looking to technology to solve the problems wrought by technology. One current embodiment of technological salvation arrives in the form of *in vitro* meat, a move from the factory to the laboratory that strives to remove the animal from the equation. The animal is no longer a "protein machine with flaws" (Pollan 2006, 219); now it is just a protein machine.

One approach we could take in evaluating these various meat-production systems would be to compare good vs. bad technological intervention, examining how technologies can corrupt or rescue meat. We could, for example, compare industrialization and its instantiations (e.g., concentrated animal-feeding operations, or CAFOs) to *in vitro* meat production, comparing their respective outcomes. Instead, I want to suggest that we are better off recalibrating the discussion. What if we looked at meat as a technology, not just the product of technology? And what if, in so doing, we came to understand technology as a means of relating to the natural world, predicated on causation and maintenance? I argue that the deeper assumptions shared by technological intervention can shed light on the ethical questions we face in making and remaking meat.

Thinking about meat as a technology in this way leads us to think about the idea of responsibility. In particular, we explore a system that produces meat and that takes responsibility for natural processes. Focusing on meat as a technology will enable us to see connections between industrialized meat production, my irresponsible attempts, and new methods of making meat meant to minimize or even eliminate moral problems. I also suggest that the visibility of such connections can enrich our ethical discussions about meat. Focusing on responsibility, refracted through the lens of technology, allows us to more fully recognize the stakes involved, and what kinds of new responsibilities we are taking on as we look for solutions to the problems of meat. By treating meat as a technology, we can map the stakes involved in the contestations over animals within our system, especially as we move to make meat "better."

Technology as a Problem for Meat

The idea that technology has created a problem for meat, especially in the present-day context, is not a new or especially controversial one. Many people have documented the historical development of the industrialized meat-production system (Horowitz 2006) and the consequences associated with this industrial turn for animals (Weis 2013), farmers (Philpott 2010; Novak 2012), workers (Striffler 2005; Cook 2010), communities (Stull and Broadway 2004), ecologies (Weis 2013), and consumers (Ogle 2013). Gains in efficiency achieved through the employment of technologies such as CAFOs—producing more meat more quickly with fewer inputs and fewer people—have come through exerting greater control over animals at all phases of the life cycle, and over the labor of the people associated with meat production, resulting in greater stresses at all points in the process (Imhoff 2010). Animals reach mandated weight more quickly through the production and administration of specialized diets and antibiotics. Mechanization allows for greater concentration of production and processing. Increased understanding of biology has led to increased manipulation of animal bodies to promote the speed and the type of growth. The success of this system has other costs on the consumer end, related to the wide availability of food that is cheap to buy yet manages to hide the true costs of production.

Because my own experiences have focused on chickens, I tend to organize my thinking on this topic through chicken-related (or at least chicken-adjacent) terms. The following problems (which can be abbreviated PEEEEP) are complex, interrelated, multivariate, and seemingly intractable:

Physiological The animal body resists uniformity, grows undesirable elements that are non-edible, and exhibits unprofitable behaviors. As Horowitz (2006, 131) puts it, "manipulation of animal biology sought to overcome the twin obstacles of natural growth rates and variations in size."

Ecological CAFOs generate waste and pollute local communities and ecosystems, which is part of what Weis (2013) calls the "ecological hoofprint." Globally, CAFOs contribute greenhouse gases to the atmosphere.

Economic Persistent inefficiencies in raising animals for meat, along with the vertical integration of meat-production systems, leave many farmers with much of the financial risk of growing animals, and little of the reward for it. This integration, pioneered by the chicken industry has spread to other meat-production industries (Striffler 2005).

Epidemiological Health and disease issues plague animals housed in CAFOs (Kirby 2011). Overconsumption of animal products can create long-term negative health effects in humans (Simon 2013).

Ethical and political Concerns over animal welfare lead to consumer boycotts and political action (Joy 2008).

These problems are not mutually exclusive, of course, and some solutions to particular problems (e.g., antibiotics given to animals sickened by confined environments and simple diets) exacerbate other problems (e.g., environmental overuse of antibiotics and resistant bacteria). It is important to recognize, however, that these problems emerge with, or are exacerbated by, the industrialization of meat production, which in turn is embedded within a certain sociotechnical system—a system in which consumers rely on efficient production of food for which only a few are responsible. In other words, people have become dependent upon a flawed system. The problems outlined above (PEEEEP) are threats to that system.

Growing awareness of and concern about these problems associated with making meat has led to a variety of responses. For consumers, the responses include vegetarian or vegan diets (Foer 2009; Pluhar 2010), greater awareness of behaviors meant to challenge the contexts of production (Keith 2009), and advocacy for changing the laws governing meat animals. For producers responding to market pressures or asserting ideals, the responses include minor modifications, such as those demanded from suppliers by the McDonald's fast-food chain (Michel 2012), and more fundamental changes to animal husbandry (Logsdon 2004). The latter tack involves taking the animal body as the limit of technological intervention, to which the system itself should be responsible.

Grass farming (Salatin 1995) exposes the problems of the farm/factory system, suggesting that the solution could be a move toward less intensive farming. Animals are rotated through a grazing schedule that imposes less stress on their bodies than factory farming. At present, grass farming occupies a rather "boutique-ish" market niche, with higher consumer prices, thus complementing rather than challenging the existing structure (McWilliams 2012). In this chapter, I will turn to a more radical approach that hopes to solve most, if not all, of the present problems associated with meat.

Technology as a Solution for Meat

Although the piecemeal approaches to change discussed above have value, they rely on modifications to consumers' behaviors—in diet, in budget, in time, and in political action. An approach that is typically more lauded in the age of high technology treats consumers' behavior as relatively intractable, and attempts to design better technologies that enable the same behaviors by consumers but remove the ecological, ethical, or political problems associated with factory farming. Problems are understood in the context of engineering rather than sociotechnical transformation. We see this approach in a host of problem areas, including hybrid automobiles and no-carb foods. Consumers' actions can stay relatively the same, while all the problems are engineered away. In the context of meat, one potential solution currently receiving a large amount of press is *in vitro* meat (Wolfson 2002; Jones 2010; Specter 2011; Hopkins and Dacey 2008; Pluhar 2010).[1]

In vitro meat technologies are aimed at using stem-cell techniques to culture meat protein in a suitable medium,[2] without the need for the entire animal. Muscle cells are grown directly, either in thin sheets or on an edible scaffold, are fed with a nutrient serum, are stimulated to simulate exercise, and are harvested as protein. In her 2007 book *Culturing Life*, Hannah Landecker examines how emerging techniques in biotechnology change "what it is to be biological." Taking cell cultures out of organic bodies makes much more control and manipulation of cells practicable. The biological becomes plastic. This plasticity works to maximize the number of non-biological elements of the technological system that can be maintained, and to modify the problems out of the animal. These broader alterations in techniques that can be used with biological matter, and in attitudes toward it, have been applied to the production of meat.

Many of the technology's developers and advocates believe it to solve the myriad problems associated with industrial meat production. A US group called New Harvest promotes *in vitro* meat as solving the following problems (Edelman et al. 2004):

- Composition—can control fat content, adding back in only specific fats, in specific quantities, deemed desirable by current dietary standards;
- Disease control—reduce unsanitary conditions by eliminating waste generated by animals or ecological systems, replaced by cells in labs;
- Efficiency—"Inedible animal structures (bones, respiratory system, digestive system, skin, and the nervous system) need not be grown";

• Exotic meats (rare and extinct)—the process could be applied to any starter muscle culture, thus allowing for any type of protein cultivation;
• Reduction of animal use—cell lines could be cultivated from a single animal.

These potential solutions rely on the idea that the process of growing meat without the animal can reach peak efficiency and efficacy in large part because the entire process is controlled. Nothing undesirable is in it; we can entirely control the nature of the protein. More importantly, this form of production eliminates the uncertainty—the risk—of natural foods. According to a spokesperson for a processing facility (quoted on page 97 of Pollan 2006), "natural ingredients are a 'wild mixture of substances created by plants and animals for completely non-food purposes—their survival and reproduction.' These dubious substances 'come to be consumed by humans at their own risk.'" *In vitro* meat eliminates the environment in which animals would normally live. It eliminates wasteful translation of energy into non-consumable elements. It eliminates moral prohibitions against eating exotic animals, or eating animals at all, because it eliminates the animal. And it eliminates the need for consumers to change—except, of course, by becoming willing to eat meat from a laboratory.

The potential of *in vitro* meat has attracted a lot of support from a variety of entities and individuals concerned about the problems associated with meat production and human overconsumption. It is significant that animal rights advocates and critics of industrialized meat production accept this technological pursuit as a reasonable (Pluhar 2010; Singer 2013) or even necessary (Hopkins and Dacey 2008; Schonwald 2009) attempt to solve the problems of meat. Some are enthusiastic supporters (Saletan 2006); others express their support as a kind of realism, preferring that more people would become vegetarians, but seeing better meat as a better bet (Deych 2005). People for the Ethical Treatment of Animals (PETA) went so far as to announce a mostly symbolic prize of a million dollars for the first lab to create an affordable and marketable *in vitro* chicken breast. (See http://www.peta.org/features/vitro-meat-contest/.) All proponents see *in vitro* meat's prime virtue as that it might allow people to remain carnivorous while eliminating the associated evils.

Such enthusiasms notwithstanding, there are barriers to making *in vitro* meat a viable source of meat protein for the general public. There are three problematic aspects of the production process:

Price At present, protein production is very expensive. In 2013, in London, a hamburger made of *in vitro* meat was taste-tested in an effort to demonstrate proof of concept and edibility. That single hamburger, funded by Sergey Brin of Google, was purported to cost $325,000 (Fountain 2013a).

Texture That same hamburger was described as having the mouth feel of cake (Fountain 2013b). This relates to a more fundamental biological problem. To more closely mimic a typical muscle, researchers have to generate or replace blood vessels, connecting tissue, and a suitable, edible, three-dimensional scaffold.

Acceptance There are groups that seem primed for acceptance of *in vitro* meat: the less finicky, those on restrictive diets, and members of the technological or gastronomical avant-garde. However, a large number of people remain very skeptical toward it (Cannavò 2010); acceptance of it follows from a kind of functional equivalence. "What makes meat 'real,'" Hopkins and Dacey (2008) write, "is its constituent substance, not its mode of production."

These are barriers for engineers and for marketers. Efforts such as the taste test are part of the campaign to pave the way for acceptability. Importantly, the ethical questions surrounding eating meat are not so much engaged with as eliminated. People are not asked to confront the ethics of eating meat—whether killing animals for food is wrong, or whether factory farming is harmful for the environment. Instead, meat is presented as a finished product that leaves current values unexamined.

Meat as Technology, Technology as Responsibility

I want to complicate the idea that *in vitro* meat solves the ethical dilemmas of technologically mediated meat production, that it simply silences the PEEEEP. I want to do this by collapsing the conceptual distance between technology and meat. Rather than thinking about how technology has been used to make meat, we can think of meat as a technology. In order to proceed, I have some work to do. Why does it make sense to conceptualize meat as technology, and how should we, in turn, conceptualize technology? In this section, I want to explore meat as a result of a human process, and technology itself as a specific kind of process—not a collection of objects as much as a series of relationships—between ourselves and the natural world.

At first blush, one might find it odd to call meat a technology. As Edgerton (2007) notes, we tend to associate technology with invention and innovation. We tend to overlook old, mundane, and ubiquitous technologies, and few things are older, more mundane, or more ubiquitous than meat. Yet if we think of technology as the product of human intervention in the natural world, then we can see that meat is, in fact, the product of active (and increasing) human intervention. It is also an object—something not simply reducible to the animal. Animals, of course, are a necessary part of the process (at least for now, pending advances in *in vitro* meat), but so are ecosystems and, in most present-day encounters, distribution systems. We can claim that meat is a technology without expanding that claim to animals themselves.

In fact, we typically do not confront animals directly when dealing with meat, nor do we need to conflate the two. We confront a disarticulated substance, something abstracted from its origins. Present-day humans exist in an era that Bulliet (2005) has dubbed "post-domesticity," referring to the state of affairs typified by fewer people having direct contact with farm and working animals, and by animals having become mere abstractions and subjects of anthropomorphic projections. The sense of remove is not just historical; it is also philosophical and cultural. Fiddes (1991) notes that humans tend to become discomfited by reminders of meat's animal origins. Or, as Vailles (1994) puts it, "we demand an ellipsis between animal and meat," accomplished in part through a distance between where humans live and where animals are slaughtered. For the modern eater, meat, like most other foods, is something to consume as a means to something else, not a focal point. As Gene Kahn, the founder of Cascadian Farm, told Michael Pollan (2006), "This is just lunch for most people. *Just lunch.* We can call it sacred, we can talk about communion, but it's just lunch."

This objectification and focus on use gives meat characteristics similar to most technological objects. Latour (2002) discusses technologies confronted as objects in what he calls the "technological mode of existence," wherein humans strive to employ technologies (and adapt in order to employ them). These objects afford us new possibilities. Some man-made objects (for example, shoes, doors, and cups) may no longer be seen as technologies, but they have altered the world we inhabit. Similarly, we use meat as a *means* to be in the world, it alters our world, and the fact that it does not typically present itself as a technology only makes it more like the countless other ordinary and "invisible" technologies we routinely encounter.

But technology does not just refer to objects. It also refers to systems and to relationships. Latour (2002) complements the technological mode with what he calls the "moral mode"—what we are forced to reckon with when a technological object ceases to function, for example. When the power goes out, or the networks go down, suddenly, a device becomes a problem. We might then realize how we are connected to routers and support staff, to power lines and power plants, even to manufacturing facilities and working conditions. Similarly, our view of meat can change when it becomes a part of a menu of choices, as something perhaps to avoid, or when it reveals one's values. Meat can be a means to accomplish something, but only to the extent that it is also an end of a very complex chain of events—animals, people, machines, contexts, and ecologies. How that chain of events gets arranged is an important question that relates to our orientation to the natural world and to the responsibilities we are willing to accept.

Thus, thinking about meat as a technology illustrates how we have made meat by remaking both the natural and social worlds. The presence of meat as a stable, reliable food, which allows humans to focus on other tasks, is itself made possible by transformed, simplified, and controlled systems that contain animals, humans, and ecologies. This mode of analysis helps capture how we have become responsible for causing and maintaining these production systems. That is, the process involves a triple transformation—of the dead animal into meat, of the living animal into a more efficient protein producer, and of ecological and human support systems. Each of these transformations is a product of causation and maintenance, and represents a kind of responsibility.

Let's consider cause first. One way of exploring technology as responsibility is in terms of causation—of bringing some thing into being as that specific thing. There is a lot of complex philosophical history dealing with the idea of bringing something into being; it exceeds the scope of this chapter. Instead, I want to sketch the legitimacy and limitations of Martin Heidegger's idea that technology is always related to responsibility. Heidegger (1977) conceptualizes technology as a basic understanding of the world. He revisits Aristotle's four causes, which emphasize what is necessary to bring something into the world or to cause something to be the way it is. The two causes most relevant to the present discussion are the material cause and the efficient cause. The material cause is the, well, material—the stuff with which we work. The efficient cause is us, the maker. We have to work with the world in order to accomplish our goals. Anyone who has ever thrown clay on a pottery wheel can attest to this. The point I want to make is that understanding technology is a way of accounting for a shared responsibility of bringing something into the world, and thus what we need to be able to evaluate is not whether we are responsible, but rather how that responsibility is configured. How specific are our demands on the world? How much do we require the world to adhere to our strict requirements?

Causation is a shared proposition. Humans cannot simply impose their will on the natural world but can work within the possibilities the world affords, and seek to create something actual. This is what happens with meat. Humans have attempted to make meat better by bringing into being better protein machines that can be tinkered with and adjusted, with all aspects of the process controlled and, for the most part, simplified. Chickens were, according to Boyd (2001: 636), "in the vanguard of animal improvement efforts." (Also see Ogle 201.) The origins of today's "meat-type" chicken (meatier overall than yesterday's chicken, and with more white meat) can be traced, in part, to the A&P food stores' "Chicken of Tomorrow" contest of the 1940s (Boyd 2001; Horowitz 2006). Efforts to breed chickens for certain behaviors and for the quality of their meat resulted in the creation of new types of birds that have helped to transform the United States into more of a chicken-eating culture.

What we have caused— the kind of meat-production system we have become responsible for—requires that we alter animals and ecosystems to produce meat that is fully disarticulated from the context of its creation. The problems that were discussed earlier in this chapter emerge because of difficulties in exerting control and enforcing simplification to manage animals and production systems. Qualities that chickens possess which are not directly related to protein production—e.g., desire for space and pecking order—are now encountered as problems to be solved. A systemic emphasis on efficiency leads to pressures for birds to be standardized in shape and behavior in order to transcend the "twin obstacles" of how slowly animals grow and how variably they are shaped Horowitz (2006, 131). This leads to interventions that regenerate bodies (e.g., accelerated growth rates) and organize space and time for those bodies in ways that support that value (e.g., through concentrating animals and employing specialized diets), but generate other problems (e.g., animals' bodies cannot support the additional muscle mass). These types of chickens become simplified versions of what is possible, emphasizing meat production over other traits, and human-shaped ecologies are created to support such production. The nature of the chicken has become less robust, simplified, and in need of control.

These simplifications include the environments that surround these versions of chickens. As Striffler (2005) points out, what had been a way for rural homesteads to produce supplemental protein and income in the early part of the twentieth century in the United States had become, by the late 1980s, a largely consolidated, vertically integrated system controlling all parts of the chicken life cycle. For example, growers would contract with companies who promised to buy chickens, but only if those birds were purchased from particular hatcheries, fed particular feed on particular schedules, grown to particular specifications, and harvested at particular times. Meanwhile, consumers came to know chicken less as a whole bird, and more as a value-added, processed entity consisting of tenders, nuggets, and patties (Horowitz 2006).

Now let's turn to maintenance. Responsibility also takes the shape of ensuring the sustainability of objects and systems. These technological systems must be maintained in order to ensure that a course of action stays stable. It is not enough to build a power plant; it has to be kept running. A classic example of this form of maintenance comes from McPhee's (1989) description of the Old River Control Structure, separating the Mississippi and Atchafalaya rivers. That structure holds the Mississippi River to its current course, and because so much depends upon it the Army Corps of Engineers strives to maintain a system that supports an object that is constantly under threat of collapsing. Humans express their intentions on the natural world through and with technologies. How much do we want to assert our intentions onto the world? It is important to see this question as not having an exclusively negative valence. We quite reasonably desire to have an effect on the world so that we may have some sense of stability. Thus, we need to realize that the more we assert our intentions, the more demands we make on the world, the more the burden of responsibility shifts to us to maintain the viability of the world that we created.

In terms of meat, the more we require animals to be highly efficient protein machines above all else, the more human systems become responsible for those animals—for feeding them, keeping them alive, and keeping them healthy. These animals don't eat as much as they are fed. They don't reproduce as much as they are bred. Their lives are maintained by human-driven ecosystems. A dependence has developed that is bred into the very beings we keep—as Pluhar (2010, 459) notes, "by changing [domesticated animals'] evolutionary paths to render them beneficial to us, we have incurred obligations of assistance." The system of meat production is maintained through active and continual intervention to ensure that the process proceeds apace. Humans cannot just bring meat into being; they must also seek to maintain the viability of the system that produces it. The system of meat production is maintained through active and continual intervention to ensure that the process proceeds. Humans do not just bring meat into being; they must also maintain its viability.

The Future of Meat

Now we are prepared to assess *in vitro* meat, the purported solution to so many of the problems of industrialized meat. This technology is offered as a potential solution to these problems, as long as the technical problems (texture and price) and the cultural problems (acceptance) can be overcome. But within the existing framework of assumptions, *in vitro* meat looks like the holy grail of the food world. It solves all of the problems outlined above while asking little of the consumer. But how does it understand the natural world? What kinds of responsibilities does it accept or demand?

For *in vitro* meat, the kind of nature we want is a direct expression of our intentions. Its chief virtues are how malleable the components are, how simplified the output is, and how controllable the process is. In fact, we end up doubling down on simplification, which was identified as the source of problems in the previous industrialized system. Or, rather, simplification was the desire, but animal bodies kept complicating things. We eliminate everything we don't want—from waste to nervous systems to behavior. Winston Churchill expressed the desire well before it was possible, in a 1931 essay titled "50 Years Hence": "We shall escape the absurdity of growing a whole chicken in order to eat the breast or wing, by growing these parts separately under a suitable medium." He was off by about thirty years, but the sentiment is familiar. By building protein up from cells, we produce only what we want.

In vitro meat also doubles down on control by building up only what is desired, and introducing only what is useful. An extreme level of control is necessary for the very viability of this technology. The productive environment has to be kept pure, lest unwanted substances take advantage of such a fecund context and corrupt the process. These bioreactors are so fertile that, according to one researcher, "We need completely sterile conditions. If you accidently add a single bacterium to a flask, it will be full in one day" (Specter 2011, 37). Researchers must also be able to control how the cells divide, so they don't become what he calls "genetic miscreants" (ibid.). Supporters of *in vitro* meat in the popular press and animal rights groups tout it as this exact kind of solution—one that can limit meat to only exactly what we want it to be—in composition, and in nature.

To achieve this control, the maintenance done must be increased. The work of cultivating muscle cells into what resembles a muscle requires a high level of understanding. The body can be removed from the process only to the extent that it is understood what a muscle is, how it works, how it comes into its specific form, what role the body plays in giving shape to the muscle, and how to manage and cultivate cells. Previously, when meat was produced in the context of an entire animal, one did not have to fully understand how that process worked to achieve a particular result. Granted, people have always sought greater understanding in order to exert more control, but that knowledge was neither as extensive nor necessary to create muscle tissue in animals. The equation changes in *in vitro* systems. We have taken on the responsibility of understanding and replacing both the animal and its environment necessary to produce protein.

Such responsibility also involves work. We not only seek to understand the functional necessities, but also to provide them. To keep the *in vitro* technology viable, the interactions caused by an animal's body moving itself through space must be replaced (electrical stimulation, flexed sheets). Nutrients must be delivered and wastes removed (nutrient baths, filtering, scaffolding). Tissue must be protected from contaminants (sterile environments).

Thus, the possibility of meat made through this technique presents a lot of potential benefits, highlighted by greater control over the environment and the animal body, such that meat becomes precisely more of what we want it to be: not reliant upon a sentient body, not contaminated by bodily functions or less controlled environments, and tailored to an idealized, healthy human body. These benefits are achieved through an orientation to nature that requires even greater responsibility for it, as simplified and maintained by our technical and social systems.

Conclusion

This increasing responsibility reflects an historical tendency to exert more control over all facets of the meat-production system, including the lives of food animals. This control is made possible through an increasing responsibility for simplified versions of these animals that can only survive within highly managed ecosystems—for example, those in the factory or, with *in vitro*, those in the laboratory. Increasing simplification and control are pursued in order to leave the role of the consumer as unchanged as is possible. *In vitro* meat solves the problems associated with simplification and control of animals associated with industry by asserting those imperatives more fully—furthering their logic. It achieves this through an orientation that sees the natural world as essentially plastic, manipulable, and changeable. Consequently, the burden of responsibility shifts even more fully onto human systems in a world made in our own image.

The more we exert controls over that animal life, through the employment of knowledge and the application of work, the more that life becomes simply an extension of human designs. The animals fade into the background, rather than being brought to the forefront of our confrontations with meat, with nature, and with our ethical priorities. For many, this is the appeal of the technology. This process of materially erasing the animal from meat can only occur in a nature that does human bidding—a nature that fully reflects human intention. Rather than confront the ethical questions of engaging animals, humans, and environments in the context of meat, we turn those questions into engineering ones. This process creates animals that become something fully realizable in the lab and fully replaceable in the production of protein. Lewis Lapham, bemoaning the current place of animals within food systems and cultural awareness, writes:

> Out of sight and out of mind, the chicken, the pig, and the cow lost their licenses to teach. The modern industrial society emerging into the twentieth century transformed them into products and commodities, swept up in the tide of economic and scientific progress otherwise known as the conquest of nature. Animals acquired the identities issued to them by man, became labels marketed by a frozen-food or meat-packing company, retaining only those portions of their value that fit the formula of research tool or cultural symbol—circus or zoo exhibit, corporate logo or Hollywood cartoon, active ingredient in farm-fresh salmon or genetically modified beef. (2013, 17)

As the animal loses its otherness, it becomes only what we want it to be. This process is furthered, rather than alleviated, in the context of *in vitro* meat technologies.

This kind of "impoverishment of experience" (Warkentin 2006, 100) created through the elimination of the otherness of animals also reduces our own needs to confront difficult ethical questions. Rather than confront our preconceptions and our practices, we expect the world to be modified around them. This technological solution works only to the extent that we manage the increased engineering complexity associated with a less complex ethical landscape for the user. And in this process humans can be understood as becoming simplified as well, ethically stunted individuals less capable of morally responding to current problems.

Thus, in assessing any technology, it is important to examine not only its technical attributes but also its worldview. It is important, that is, to imagine a future that not only contains *in vitro* meat, but the attitudes and responsibilities that make such a technology possible. These include the view of the natural world as plastic, the virtue of control that becomes a necessity, the increasing responsibility human systems take on in maintaining the technologies they create, and the decreased complexity associated with the user of those technologies. The central ethical question is this: What kind of world, what kind of human, what kind of nature do we anticipate in our technological designs? It is a question both of capability and of desire.

Notes

1. Though the present approach to *in vitro* meat reflects more recent advances in tissue culture and experiments with stem cells, the idea of growing meat without the animal has a longer history. Scientific experimentation with chicken heart tissue conducted by Alexi Carrell (Jiang 2012) led to boosterism (Churchill 1931) and trepidation (Obler 1937; Pohl and Kornbluth 1969).

2. That medium currently involves fetal bovine serum, which does not escape the need for animal bodies. Efforts to engineer a viable, vegetable based derivate for that medium are under way.

References

Boyd, W. 2001. Making meat: Science, technology, and American poultry production. *Science and Culture* 42 (4): 631–664.

Bulliet, R. W. 2005. *Hunters, Herders, and Hamburgers: The Past and Future of Human-Animal Relationships*. Columbia University Press.

Cannavò, P. 2010. Listening to the "yuck factor": Why in-vitro meat may be too much to digest, Presented at annual meeting of American Political Science Association, Washington.

Churchill, W. 1931. Fifty Years Hence. *Strand Magazine* (http://teachingamericanhistory.org/library/document/fifty-years-hence/).

Cook, C. 2010. Sliced and diced: The labor you eat. In *The CAFO Reader: The Tragedy of Industrial Animal Factories*, ed. D. Imhoff. Watershed Media.

Deych, R. 2005. How one vegan views in-vitro meat. http://www.rrrina.com/invitro_meat.htm

Edelman, P. E., D. C. McFarland, V. A. Mironov, and J. G. Matheny. 2004. In vitro cultured meat production. http://www.new-harvest.org/img/files/Invitro.pdf

Edgerton, D. 2007. *The Shock of the Old: Technology and Global History since 1900*. Oxford University Press.

Fiddes, N. 1991. *Meat: A Natural Symbol*. Routledge.

Foer, J. S. 2009. *Eating Animals*. Little, Brown.

Fountain, H. 2013a. Building a $325,000 burger. *New York Times*, May 14.

Fountain, H. 2013b. Frying up a lab-grown hamburger. *New York Times*, August 5.

Galusky, W. 2010. Playing chicken: Technologies of domestication, food, and self. *Science as Culture* 19 (1): 15–35.

Heidegger, M. 1977. *The Question Concerning Technology, and Other Essays*. Harper & Row.

Hopkins, P. D., and A. Dacey. 2008. Vegetarian meat: Could technology save animals and satisfy meat eaters? *Journal of Agricultural & Environmental Ethics* 21 (6): 579–596.

Horowitz, R. 2006. *Putting Meat on the American Table: Taste, Technology, Transformation*. Johns Hopkins University Press.

Imhoff, D. 2010. *The CAFO Reader: The Tragedy of Industrial Animal Factories*. Watershed Media.

Jiang, L. 2012. Alexis Carrel's immortal chick heart tissue cultures (1912–1946). In *Embryo Project Encyclopedia* (http://embryo.asu.edu/handle/10776/3937).

Jones, N. 2010. A taste of things to come? *Nature* 468: 752–753.

Joy, M. 2008. *Strategic Action for Animals: A Handbook on Strategic Movement Building, Organizing, and Activism for Animal Liberation*. Lantern Books.

Keith, L. 2009. *The Vegetarian Myth: Food, Justice, and Sustainability*. Flashpoint.

Kirby, D. 2011. *Animal Factory: The Looming Threat of Industrial Pig, Dairy, and Poultry Farms to Humans and the Environment*. St. Martin's Press.

Landacker, H. 2007. *Culturing Life: How Cells Became Technologies*. Harvard University Press.

Lapham, L. H. 2013. Man and beast. *Lapham's Quarterly* 6 (2): 13–21.

Latour, B. 2002. Morality and technology: The end of the means. *Theory, Culture & Society* 19 (5/6): 247–260.

Logsdon, G. 2004. *All Flesh Is Grass: The Pleasures and Promises of Pasture Farming*. Swallow.

McPhee, J. 1989. *The Control of Nature*. Farrar, Straus and Giroux.

McWilliams, J. W. 2012. The myth of sustainable meat. *New York Times*, April 12.

Michel, M. 2012. McDonald's US pledges to phase out sow stalls. GlobalMeatNews.com (http://www.globalmeatnews.com/Livestock/McDonald-s-US-pledges-to-phase-out-sow-stalls).

Novak, J. 2012. Discipline and distancing: Confined pigs in the factory farm gulag. In *Animals and the Human Imagination: A Companion to Animal Studies*, ed. A. Gross and A. Vallely. Columbia University Press.

Obler, A. 1937. Chicken heart [radio series episode]. Retrieved from https://archive.org/download/LightsOutoldTimeRadio/LightsOut-1937-03-10ChickenHeart.mp3.

Ogle, M. 2013. *In Meat We Trust: An Unexpected History of Carnivore America*. Houghton Mifflin Harcourt.

Philpott, T. 2010. Squeezed to the last drop: The loss of family farms. In *The CAFO Reader: The Tragedy of Industrial Animal Factories*, ed. D. Imhoff. Watershed Media.

Pluhar, E. 2010. Meat and morality: Alternatives to factory farming. *Journal of Agricultural & Environmental Ethics* 23 (5): 455–468.

Pohl, F., and C. M. Kornbluth. 1969. *The Space Merchants*. Walker.

Pollan, M. 2006. *The Omnivore's Dilemma: A Natural History of Four Meals*. Penguin.

Salatin, J. 1995. *Salad Bar Beef*. Polyface.

Saletan, W. 2006. The conscience of a carnivore. http://www.slate.com/id/2142547/

Schonwald, J. 2009. Future fillet. *University of Chicago Magazine* (http://magazine.uchicago.edu/0906/features/future_fillet.shtml).

Simon, D. R. 2013. *Meatonomics: How the Rigged Economics of Meat and Dairy Make You Consume Too Much—and How to Eat Better, Live Longer, and Spend Smarter*. Conari.

Singer, P. 2013. The world's first cruelty-free hamburger. *The Guardian* (http://www.theguardian.com/commentisfree/2013/aug/05/worlds-first-cruelty-free-hamburger).

Specter, M. 2011. Test-tube burgers. *The New Yorker*, May 23: 32–38.

Striffler, S. 2005. *Chicken: The Dangerous Transformation of America's Favorite Food*. Yale University Press.

Stull, D. D., and M. J. Broadway. 2004. *Slaughterhouse Blues: The Meat and Poultry Industry in North America*. Thomson/Wadsworth.

Vailles, N. 1994. *Animal to Edible*. Cambridge University Press.

Warkentin, T. 2006. Dis/integrating animals: Ethical dimensions of the genetic engineering of animals for human consumption. *AI & Society* 20 (1): 82–102.

Weis, A. 2013. *The Ecological Hoofprint*. Zed Books.

Wolfson, W. 2002. Lab-grown steaks nearing the menu. *New Scientist* (http://www.newscientist.com/article/dn3208-labgrown-steaks-nearing-the-menu.html).

Contributors

Braden R. Allenby is the Lincoln Professor of Engineering and Ethics and Professor of Civil, Environmental and Sustainable Engineering, and of Law, at Arizona State University. He is the founding director of the Center for Earth Systems Engineering and Management, and the founding chair of the Consortium for Emerging Technologies, Military Operations, and National Security, at ASU. His areas of expertise include industrial ecology, sustainable engineering, earth systems engineering and management, and emerging technologies. His latest books are *Industrial Ecology and Sustainable Engineering* (with Tom Graedel, 2009), *The Techno-Human Condition* (with Dan Sarewitz, 2011), *The Theory and Practice of Sustainable Engineering* (2012), and (as editor) *The Applied Ethics of Emerging Military and Security Technologies* (2015).

Raymond Anthony is Professor of Philosophy at the University of Alaska Anchorage. He specializes in ethical theory, philosophy of technology, and animal-agricultural-environmental-food ethics. He serves as ethics advisor for the American Veterinary Medical Association's Animal Welfare Committee and as a council member for Agriculture, Food and Human Values Society. He is the recipient of an inter-institutional USDA grant to develop teaching aids for education in the ethics of animal welfare. Currently he is pursuing values-aware research in global food security, sustainability, animal welfare, and climate ethics, with particular emphasis on dairy production in Southern Brazil.

Philip Brey is Professor of Philosophy of Technology and chair of the Department of Philosophy at the University of Twente. He is also director of the Centre for Philosophy of Technology and Engineering Science (CEPTES). He has published in the areas of general philosophy of technology, philosophy of science, philosophy of biomedical technology, philosophy of sustainable technology, and philosophy of information and communication technology. He co-edited, with Thomas Misa and Andrew Feenberg, *Modernity and Technology* (MIT Press, 2003); he also co-edited two special issues of the journal *Ethics and Information Technology*.

J. Baird Callicott is University Distinguished Research Professor at the University of North Texas. In 1971 he taught the world's first course in environmental ethics at the University of Wisconsin at Stevens Point. He is editor in chief of the *Encyclopedia of Environmental Ethics and Philosophy* and author or editor of a score of other books, and author of dozens of journal articles and book chapters on environmental ethics and philosophy. He has served the International Society for Environmental Ethics as president and Yale University as bioethicist-in-residence. His research focuses on theoretical environmental ethics, climate ethics, comparative environmental ethics, and the philosophy of ecology and conservation policy.

Brett Clark is Associate Professor of Sociology and Environmental Humanities at the University of Utah. He is the author of several books, including *The Tragedy of the Commodity: Oceans, Fisheries, and Aquaculture* (with Stefano B. Longo and Rebecca Clausen), *The Science and Humanism of Stephen Jay Gould* (with Richard York), *The Ecological Rift* (with John Bellamy Foster and Richard York), and *Critique of Intelligent Design: Materialism versus Creationism from Antiquity to the Present* (with John Bellamy Foster and Richard York).

Wyatt Galusky is coordinator of the Science, Technology, & Society Program and Associate Professor of Humanities at Morrisville State College. His research explores the various ways in which technologies mediate between humans and the natural world, including domesticated food animals and public engagements with science and technology. He is a recipient of grants from the New York Council for the Humanities and the US Department of Energy. His publications have appeared in the *Journal of Agricultural Ethics*, in *Science as Culture*, and in *Agricultural and Environmental Ethics*. He is currently working on a book, titled *Protein Machines*, that examines the technologies associated with producing meat.

Ryan Gunderson is an Assistant Professor of Sociology and Social Justice Studies in the Department of Sociology and Gerontology at Miami University. His research interests include environmental sociology, social theory, animal studies, the sociology of consumption, and the sociology of technology. His research projects include a sociological examination of geoengineering, deliberative environmental decision making, global environmental governance, and the relevance of classical sociological theory to the social aspects of various technologies.

Benjamin Hale is Associate Professor in the Philosophy Department and the Environmental Studies Program at the University of Colorado at Boulder. From 2006 to 2008 he was director of the Center for Values and Social Policy, and he continues active engagement with the center by co-organizing the annual Rocky Mountain Ethics (RoME) Congress with Alastair Norcross. He is co-editor, with Andrew Light, of the journal *Ethics, Policy & Environment* and vice president of the International Society for Environmental Ethics. His primary area of research focus is environmental ethics, though he maintains active interests in a wide range of ethical topics.

Clare Heyward is a Leverhulme Early Career Fellow working on the Global Justice and Geoengineering project. Before joining the University of Warwick, she was James Martin Research Fellow on the Oxford Geoengineering Programme. Her interests include global distributive justice and intergenerational justice, especially those connected to climate change.

Don Ihde is Distinguished Professor of Philosophy at the State University of New York at Stony Brook. In 1979 he wrote what is often identified as the first North American work on philosophy of technology, *Technics and Praxis*. In 2013 he received the Golden Eurydice Award. He is the author of thirteen original books and an editor of many others, including *Chasing Technoscience* (2003, edited with Evan Selinger), *Bodies in Technology* (2002);, *Expanding Hermeneutics: Visualism in Science* (1998), and *Postphenomenology* (1993). He lectures and gives seminars internationally. His books and articles have appeared in a dozen languages.

David M. Kaplan is Associate Professor in the Department of Philosophy and Religion at the University of North Texas, where he also runs the Philosophy of Food Project. His research focuses on hermeneutics, food, and technology. He is the editor of the second edition of *Readings in the Philosophy of Technology* (Rowman and Littlefield, 2009), of *Philosophy of Food* (University of California Press, 2012), and of *Reading Ricoeur* (SUNY Press, 2008), and co-editor, with Paul B. Thompson, of *The Encyclopedia of Food and Agricultural Ethics* (Springer, 2014). He is working on a manuscript on the philosophy of food titled *Food Philosophy* (to be published by the Columbia University Press).

Steve Rayner is James Martin Professor of Science and Civilization and Director of the Institute for Science, Innovation and Society (InSIS) at Oxford University, where he also co-directs the Oxford Geoengineering Programme. He has held senior research positions at two US National Laboratories and has taught at leading US universities. He has served on various US, UK, and international bodies addressing science, technology and the environment, including Britain's Royal Commission on Environmental Pollution, the Intergovernmental Panel on Climate Change, and the Royal Society's Working Group on Climate Geoengineering.

Mark Sagoff is a Senior Research Scholar at the Institute for Philosophy and Public Policy in the School of Public Affairs at the University of Maryland at College Park. He has published widely in journals of law, philosophy, and the environment. His most recent books are *The Economy of the Earth* (second edition; Cambridge University Press, 2008) and *Price, Principle, and the Environment* (Cambridge University Press, 2004). He was named a Pew Scholar in Conservation and the Environment in 1991 and was awarded a Fellowship at the Woodrow Wilson International Center for Scholars in 1998. He is a Fellow of the American Association for the Advancement of Science and of the Hastings Center.

Julian Savulescu is Uehiro Professor of Practical Ethics at the University of Oxford, a Fellow of St. Cross College, Oxford, Director of the Oxford Uehiro Centre for Practical Ethics, and Head of the Melbourne-Oxford Stem Cell Collaboration, which is devoted to examining the ethical implications of cloning and embryonic stem cell research. His areas of research include the ethics of genetics (predictive testing, behavioral genetics, enhancement, and gene therapy), medical, research, and reproductive ethics, and the ethics of geo-engineering and terra-forming.

Paul B. Thompson is W. K. Kellogg Chair in Agricultural, Food and Community Ethics in the Department of Philosophy at Michigan State University, with partial appointments in the Agricultural Economics and Resource Development departments. He previously held positions as Distinguished Professor of Philosophy and director of the Center for Food Animal Well-Being at Purdue University and as Professor of Philosophy and Agricultural Economics and director of the Center for Science and Technology Policy and Ethics at Texas A&M University. He also holds a joint position as an editor in chief of the book series *The International Library of Environmental, Agricultural and Food Ethics* and serves on the editorial boards of the journals *Agriculture and Human Values* and *Journal of Agricultural and Environmental Ethics*.

Ibo van de Poel is Associate Professor in ethics and technology at Delft University of Technology. He has done research and published works on the dynamics of technological development, on codes of conduct and professional ethics of engineers, on the moral acceptability of technological risks, on ethics in engineering design, and on ethics and responsibility in R&D networks. He has been involved in several educational innovations in the area of ethics and technology, including the development of the Web-based computer program AGORA and the first Dutch textbook on ethics and technology.

Zhang Wei is Assistant Professor of Philosophy at Central China Normal University. His research focuses on ethics of technology and environmental ethics. His recent work examines the moral valence in technologies and their potential for environmental protection. His publications have appeared in *Technē*, in *Research in Philosophy and Technology*, in *Studies in Philosophy of Science and Technology*, and in *Philosophy and Technology*.

Kyle Whyte holds the Timnick Chair in the Humanities at Michigan State University. He is Associate Professor of Philosophy and Community Sustainability, a faculty member of the Environmental Philosophy & Ethics graduate concentration, and a faculty affiliate of the American Indian Studies and Environmental Science & Policy programs. His primary research addresses moral and political issues concerning climate policy and Indigenous peoples and the ethics of cooperative relationships between Indigenous peoples and climate science organizations. He is an enrolled member of the Citizen Potawatomi Nation. His articles have appeared in the journals *Climatic Change, Sustainability Science, Environmental Justice, Synthese,* and *Hypatia*.

Index

FASKEN LEARNING RESOURCE CENTER
9000088364